MATHEMATICAL TIME EXPOSURES

By
I. J. SCHOENBERG

MATHEMATICAL TIME EXPOSURES

By
I. J. SCHOENBERG
Mathematics Research Center
University of Wisconsin, Madison

Published and distributed by
THE MATHEMATICAL ASSOCIATION OF AMERICA

Library of Congress Catalog Card Number 82-062766

ISBN 0-88385-438-4

Printed in the United States of America

Current printing (last digit)

10 9 8 7 6 5 4 3 2 1

Gratefully dedicated to the memory of

VICTOR COSTIN
ALEXANDRU MYLLER
VERA MYLLER LEBEDEFF
SIMION SANIELEVICI

my teachers at the University of Jassy, Rumania.

PREFACE

First a word about the title of this book and the 18 chapters or essays that it contains. In 1939 Hugo Steinhaus published his admirable *Mathematical Snapshots*. Although it was meant to be a popular book, due to its diversity and wide range, we all learned much from it. My present aims are roughly similar, but the pace is more leisurely; my lens is not as fast as Steinhaus's.

The subjects discussed may be vaguely described as something new, something old, and something borrowed. To begin with, no effort was made to be contemporary. In Chapter 10 we see that the 17th century problem of interpolation stimulates new ideas and methods, and Mathematics is perhaps the only science where this is possible. The subjects of Chapters 12 and 14 actually antedate the invention of the daguerreotype. Chapter 9 uses ideas of Jesse Douglas of 1960 to suggest rectilinear models for outdoor sculptures. Chapter 18 describes a new extremum property of Kepler's polyhedron known as the Stella Octangula. This chapter uses Kronecker's simplest result on Diophantine Approximations from Chapter 16, and was suggested by the work of D. König and A. Szücs of 1913 on the motions of a billiard ball within a cube.

It would seem misleading to classify papers as to level and sophistication by whether or not they appeal to concepts of Calculus. Nevertheless it so happens, without planning except for rearrangements, that the first nine essays deal with pre-Calculus subjects, while the last nine, from 10 to 18, use notions from Calculus, either implicitly or explicitly.

The advantage of a collection such as this is that if the reader does not care for one topic, perhaps the next might be more congenial. The reason is that the subjects are mostly disconnected, except the four essays on the finite Fourier series and the four on billiard ball motions. The rather tenuous interdependence among these essays is described in their introductions.

On the proper length of an essay we are reminded of Lincoln's question and answer: "How long should a person's legs be? Just long enough to reach the ground." There remains a problem: to reach the ground without becoming tedious. In Chapter 16, dealing with Kronecker's theorems, I believe we have reached the ground, but in doing that have we become tedious? Perhaps, but Fritz Lettenmeyer was certainly too brief in his remarkable paper [**2**].

I wish to thank the late Professor E. F. Beckenbach and Professor Ross Honsberger for asking me to write a book to be published by the Mathematical Association of America. The present collection originated during 1977–78 in a weekly seminar of the Mathematics Department of the United States Military Academy at West Point. I am grateful to Colonels J. W. MacNulty and J. M. Pollin for their active encouragements. Finally, my thanks are due to the Mathematics Research Center of the University of Wisconsin–Madison for its generous hospitality, and to its "student helper" Crescent Kringle for her typing of the manuscript.

I. J. SCHOENBERG

CONTENTS

CHAPTER 1

ON STEINHAUS'S FIRST MATHEMATICAL SNAPSHOT OF 1939

1. Introduction. Referring to Boltyanskii's excellent booklet [**2**], we start with the following definition. Two figures in the plane are said to be *equidecomposable* if it is possible to decompose one of them into a finite number of parts which can be rearranged to form the second figure. As an example we see that on decomposing the regular triangle of Figure 1.1(a), having all sides 2, into two right-angled triangles, these can be rearranged to form a rectangle of dimensions $1 \times \sqrt{3}$, of Figure 1.1(b).

We are concerned with polygonal figures in the plane. It is clear that a pair of equidecomposable polygons have equal areas. It is natural to ask if the *converse* holds: If two polygons have equal areas, are they equidecomposable? An affirmative answer to this question was obtained, almost simultaneously, by the Hungarian mathematician Bolyai (1832) and the German officer and amateur mathematician Gerwin (1833): *Two polygons of equal areas are equidecomposable into polygonal parts.*

It is remarkable that this result does *not* extend to polyhedra in space, as shown by Max Dehn (1902). He showed that a cube and a regular tetrahedron, both having equal volumes, are *not* equidecomposable into polyhedral parts [**2**, Chapter 5].

By the Bolyai-Gerwin theorem we know that the regular triangle with sides 2, and the square of sides $3^{1/4}$, are equidecomposable because they have the same area $3^{1/2}$. In his first snapshot (see [**5**]), Steinhaus suggests the equidecomposability of the two figures of Figure 1.2 in an interesting manner by means of a linkage.

We reproduce here this first snapshot: "From these four small boards we compose a square or an equilateral triangle, according as we turn the handle up or down." (See Figure 1.3.)

This is the entire snapshot, and Steinhaus gives no details and no references.* We turn it into our first time exposure by giving the construction of the four boards. We do this because we believe that an actual

*See the final remark at the end of this chapter.

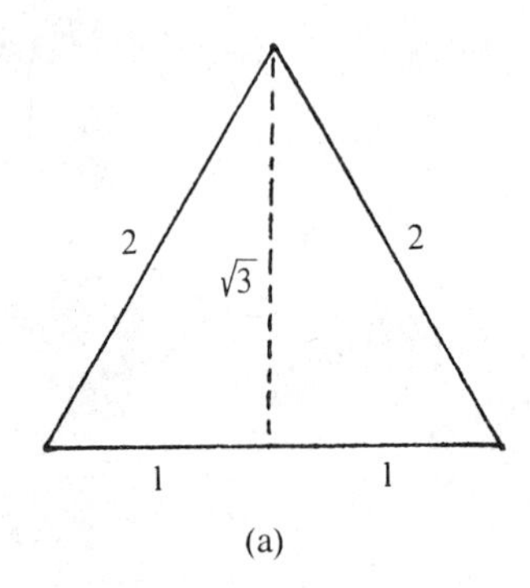

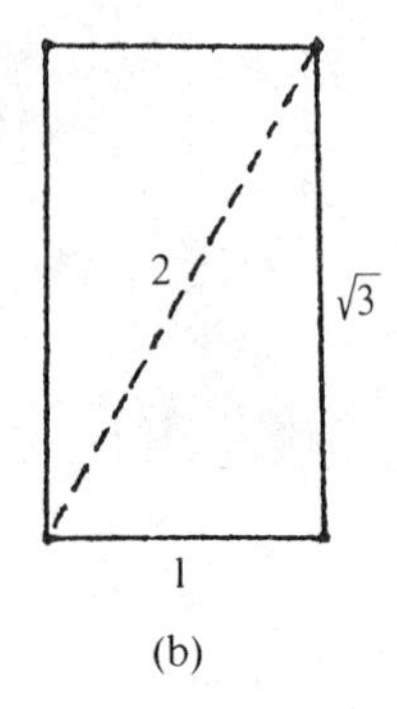

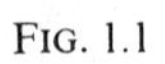
FIG. 1.1

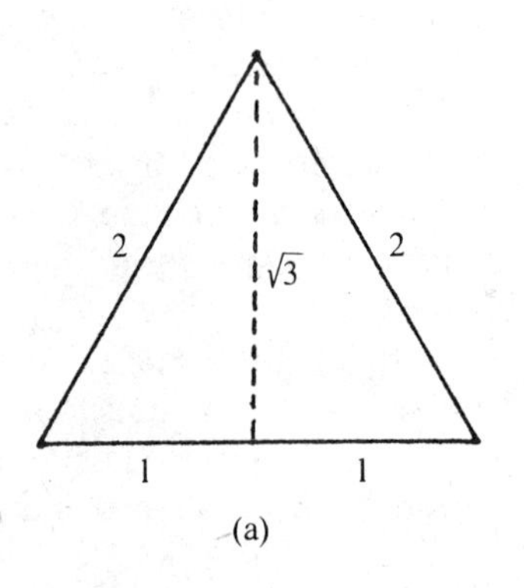

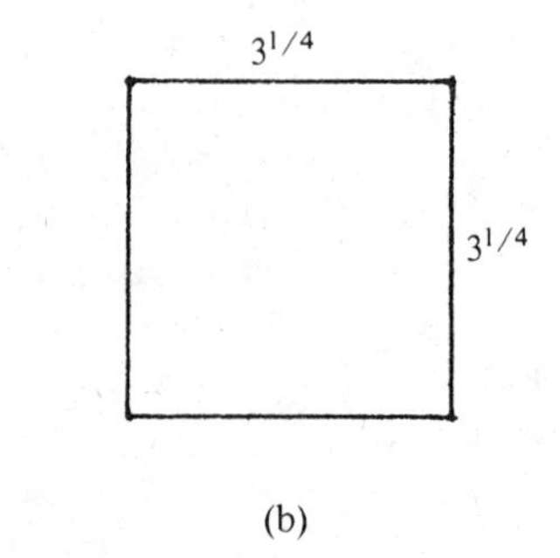

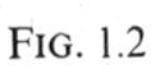
FIG. 1.2

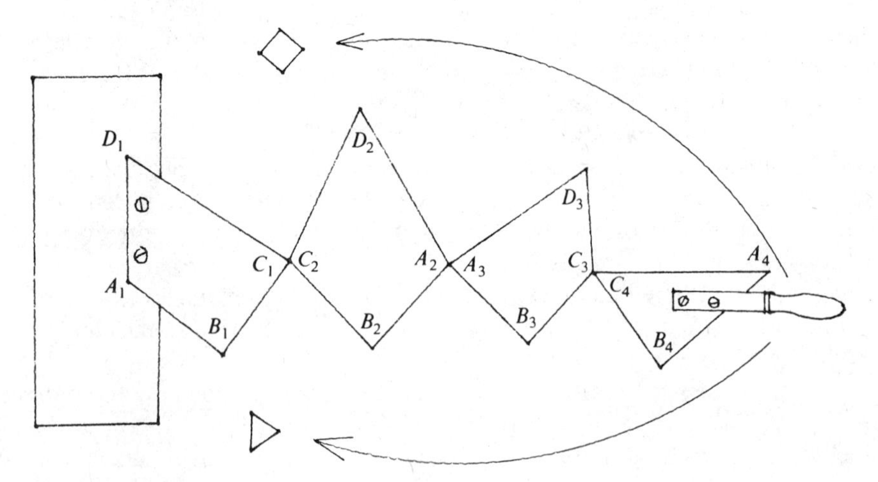

FIG. 1.3

construction of a linkage as described by Figure 1.3, made of thin wooden boards, would provide a worthwhile project for an "Open House" exhibit of some Mathematics Department.

2. The Decomposition of the Regular Triangle. Ideally, the solution of a geometric problem should contain an analysis and a synthesis. The analysis starts from the solution and derives *necessary* conditions. The synthesis shows that these conditions are also *sufficient*. For brevity we skip the analysis and give only the synthesis of the solution. We do this by describing in Figure 1.4 the dissection of the regular triangle into the four polygons.

The large regular triangle $D_1D_2D_3$, having sides 2, is dissected into three quadrilaterals and one triangle as follows:

$$A_1B_1C_1D_1, \quad A_2B_2C_2D_2, \quad A_3B_3C_3D_3, \quad \text{and} \quad A_4B_4C_4. \tag{2.1}$$

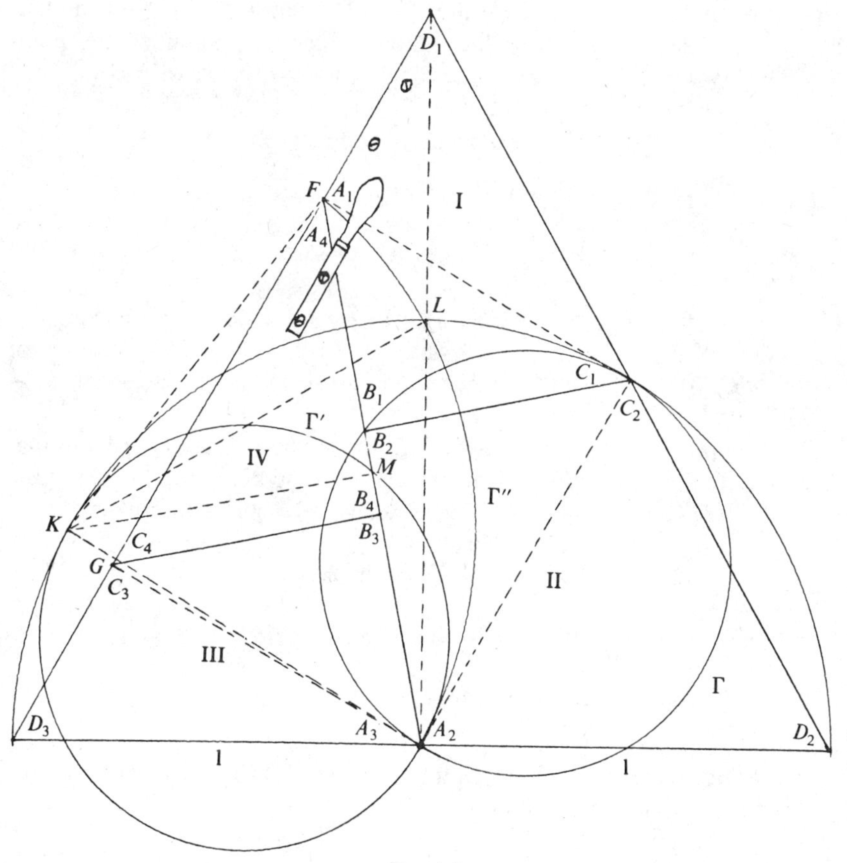

FIG. 1.4

These are supposed to be congruent to the similarly labeled "small boards" of Figure 1.3, but in Figure 1.3 these are not drawn to scale, nor in the correct shape.

The point $A_2 = A_3$ is the midpoint of D_2D_3, and $C_1 = C_2$ is the midpoint of D_1D_2. Thus,

$$A_3D_3 = A_2D_2 = 1, \qquad D_2C_2 = C_1D_1 = 1. \tag{2.2}$$

We draw the circle D_2LD_3 on the base D_2D_3 as diameter, and mark the point L of its intersection with the altitude D_1A_2. The regular triangle $A_2D_2C_2$ is now turned counterclockwise by 90° about A_2, assuming the new position A_2LK. Therefore

$$C_2A_2K = 90° \quad \text{and} \quad A_2K = KL = 1. \tag{2.3}$$

Draw the circle Γ'' having the center K and radius $KA_2 = KL = 1$, and let Γ'' intersect the side D_1D_3 in the point F. Having determined the point $F = A_1 = A_4$, we determine on D_1D_3 the point $G = C_3 = C_4$ so that

$$FG = C_2A_2 = 1. \tag{2.4}$$

It follows that

$$A_2C_2FG \text{ is a parallelogram.} \tag{2.5}$$

Finally, join F to $A_2 = A_3$ by a straight line, let $B_1 = B_2$ be the orthogonal projection of $C_1 = C_2$ onto FA_2, and let $B_3 = B_4$ be the projection of $G = C_3 = C_4$ onto FA_2. This completes our construction and determines the quadrilaterals and triangle (2.1).

We now imagine that the triangle $D_1D_2D_3$ is made of cardboard and that it is dissected into the four polygons (2.1). We also denote the figures (2.1) by I, II, III, IV, as indicated in Figure 1.4. We also join with hinges

the quadrilaterals	I	with II	at the point	$C_1 = C_2$,	
the quadrilaterals	II	with III	at the point	$A_2 = A_3$,	(2.6)
finally	III	with IV	at the point	$C_3 = C_4$.	

At this point the reader is asked to use Figure 1.4 to cut out of paper the polygons I, II, III, IV and think of them as hinged, as in (2.6).

3. Obtaining the Square. We are yet to show how the square arises by keeping I fixed while rotating II, III, IV counterclockwise. This we do as follows. We begin by orienting the triangle $D_1D_2D_3$ so that the line A_1A_2 becomes vertical. Next we perform three turning operations as follows.

(1) We turn IV about its vertex C_4 by $+180°$,
(2) we turn the union $\mathrm{III} \cup \mathrm{IV}$ about the point $A_2 = A_3$ by $+180°$,
(3) we turn $\mathrm{II} \cup \mathrm{III} \cup \mathrm{IV}$ about $C_1 = C_2$ by $+180°$, and wish *to show that the resulting figure is the square* $B_1B_2B_3B_4$ *of Figure* 1.5.

From the relations (Figure 1.4)

$$D_3A_3 = A_2D_2 = 1, \qquad D_2C_2 = C_1D_1 = 1$$

and the fact that the angles at B_1, B_2, B_3, and B_4 are all 90°, it follows that after the three rotations we obtain Figure 1.5, except that we are yet to show that we indeed get a *square*, i.e., that we have in Figure 1.5 the relations

$$B_1B_2 = B_2B_3 = B_3B_4, \tag{3.1}$$

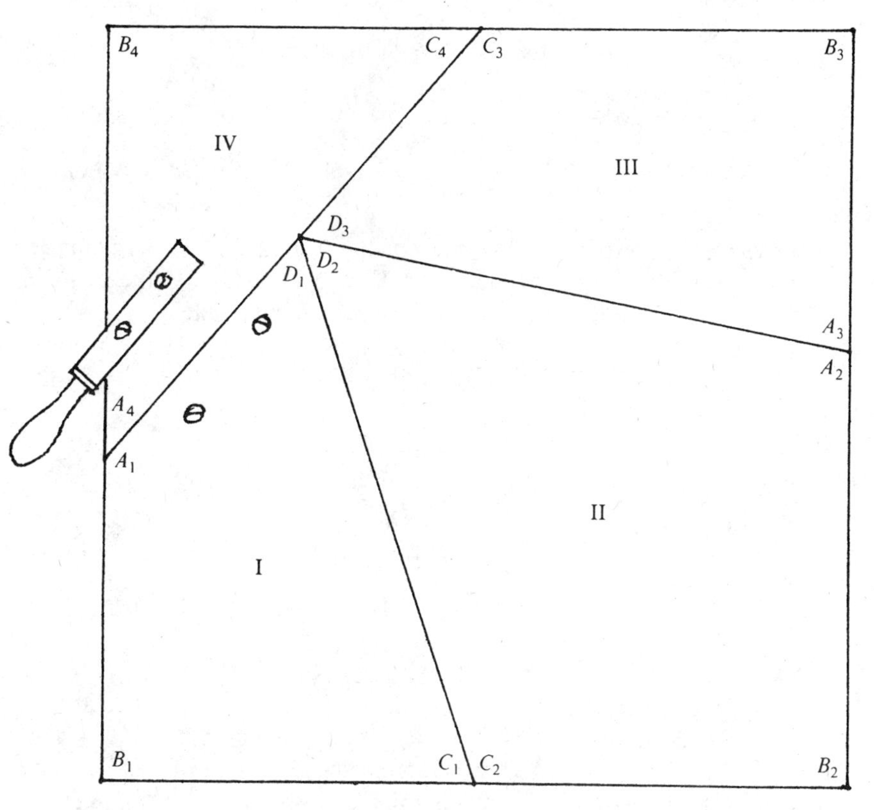

FIG. 1.5

or, explicitly, *that we have the two relations*

$$B_1C_1 + C_2B_2 = B_2A_2 + A_3B_3 \tag{3.2}$$

and

$$B_1C_1 + C_2B_2 = B_3C_3 + C_4B_4. \tag{3.3}$$

As these six segments appear in Figure 1.4, we may return to Figure 1.4.

Proof of (3.3): Since $C_2B_2 = C_1B_1$ (as they represent the same segment), and likewise $B_3C_3 = B_4C_4$, the relation (3.3) amounts to $B_3C_3 = B_1C_1$. But this is clear, because of (2.5).

Proof of (3.2): This is equivalent to

$$2 \cdot C_2B_2 = B_2A_2 + A_3B_3. \tag{3.4}$$

Since A_2C_2FG is a parallelogram we have

$$B_2A_2 + A_3B_3 = A_2B_2 + B_1A_1 = A_2F$$

and so (3.4) amounts to showing that

$$2 \cdot C_2B_2 = A_2F. \tag{3.5}$$

Let the circle Γ' intersect the line A_2F for the second time at the point M. As the center of similitude of the circles Γ' and Γ'' is clearly the point of tangency A_2, the ratio of similitude being $1:2$, we have that

$$A_2F = 2 \cdot A_2M. \tag{3.6}$$

By (3.5) and (3.6) we are left to show that

$$C_2B_2 = A_2M. \tag{3.7}$$

This is seen as follows. The two triangles

$$C_2B_2A_2 \quad \text{and} \quad A_2MK \tag{3.8}$$

are both right-angled and have their sides respectively perpendicular, i.e., $C_2B_2 \perp A_2M$ a.s.f. It follows that the triangles are similar. Since $A_2C_2 = KA_2$ $(=1)$, it follows that the triangles (3.8) are congruent. This proves (3.7) and concludes our discussion of Steinhaus's first snapshot.

Problems

1. Figure 1.6 shows a regular 12-gon of radius 1 inscribed in a square of side 2. The dissection of the 12-gon into triangles is obtained by drawing the regular star-shaped 12-gon $A_1A_2\ldots A_{12}$.

(i) Show that the 2×2 square $BCDE$ is decomposed into 16 congruent regular triangles T and 32 congruent isosceles triangles I.

(ii) Show by decomposition into T's and I's that the 12-gon is equidecomposable to the union of the three squares OA_1BA_4, OA_4CA_7, and OA_7DA_{10}. It follows that the area of the 12-gon is 3.

The determination of the area of the 12-gon given above was found in 1898 by the Hungarian mathematician Josef Kürschak and we call Figure 1.6 *Kürschak's Tile.*

2. The first problem in the 1977 International Mathematical Olympiad held at Belgrade, Yugoslavia, is as follows:

"The four equilateral triangles ABK, BCL, CDM, DAN are constructed inside the square $ABCD$ (Figure 1.7). Prove that the midpoints of the four segments KL, LM, MN, NK, and the midpoints of the eight segments AK, AN, BL, BK, CM, CL, DN, DM, are the twelve vertices of a regular dodecagon."

Prove this statement by using the Kürschak tile of Figure 1.6.

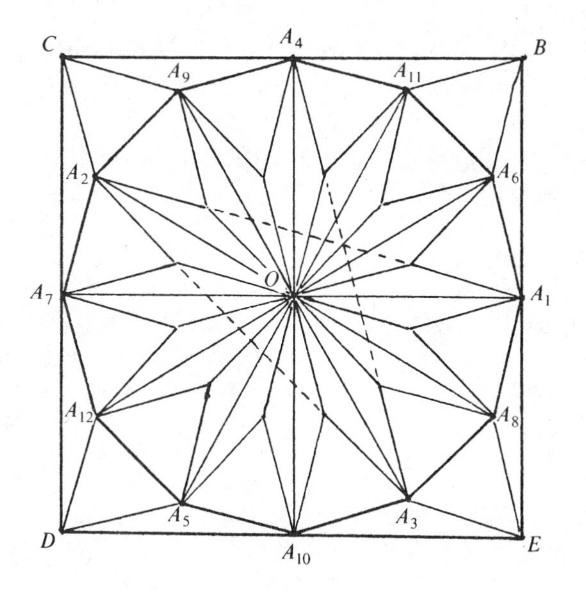

FIG. 1.6

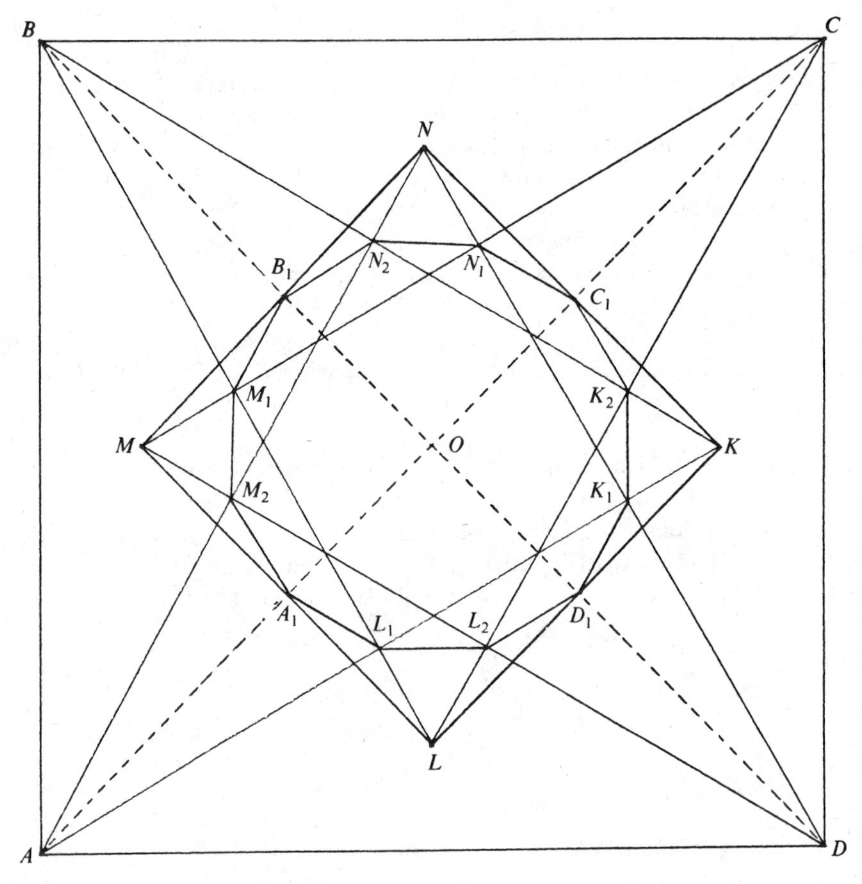

FIG. 1.7

(*Hint*: The secret is to *invert* the problem. We start with the Kürschak tile $KLMN$ of Figure 1.7 and its regular 12-gon $A_1L_1L_2 \ldots B_1M_1M_2$. To obtain the entire Figure 1.7 we extend the sides KK_1 and KK_2 of the regular triangle KK_1K_2, and do the same with the sides of the other three triangles LL_1L_2, MM_1M_2, NN_1N_2. The intersections of these extensions define the points $ABCD$. By the fact that a 90° rotation about O leaves the figure invariant, it follows that $ABCD$ is a square. Convince yourself that we have the complete figure as required by the Olympiad problem. There remains to see that the $4+8=12$ midpoints of the problem coincide with the vertices of the dodecagon.)

Final remarks. I owe to M. S. Klamkin the reference **[3]** where the reader will find the equidecomposition problem of this MTE* and many others. Kürschak's Tile suggested to me the theme of my paper **[4]**.

---

*Mathematical Time Exposure

References

[1] G. L. Alexanderson and Kenneth Seydel, *Kürschak's Tile*, Math. Gaz., (1978) 192–195.

[2] V. G. Boltyanskii, *Equivalent and Equidecomposable Figures*, D. C. Heath, Boston, 1963.

[3] H. Lindgren, *Recreational Problems in Geometric Dissections and How to Solve Them*, Dover, New York, 1972.

[4] I. J. Schoenberg, *Approximating lengths, areas and volumes by polygons and polyhedra*, Delta, 7 (1977) 32–46.

[5] H. Steinhaus, *Mathematical Snapshots*, G. E. Stechert, New York, 1939.

CHAPTER 2

ON A PROBLEM OF STEINHAUS AND SIERPIŃSKI ON LATTICE POINTS

1. Introduction. In the present MTE we start with analytical geometry in the plane, but soon pass to geometry in n dimensions. Historically this was a slow process extending over two centuries. Oversimplifying the historical development, we may mention the names of R. Descartes (two dimensions, 17th century), A. C. Clairaut and L. Euler (three dimensions, 18th century), A. Cayley and H. Grassmann (n dimensions, 19th century). For historical details we recommend the superb little book by Dirk Struik [**6**]. Why was the progress from two to three and higher dimensions so slow? Perhaps because geometers were so busy at each stage gathering the rich harvest of results and applications they had no time to pass to the next stage.

We will freely use irrational numbers, the chief glory of Greek geometry, and some 19th century generalizations associated with the name of L. Kronecker (see Chapter 23, entitled "Kronecker's Theorem," in [**1**]). These generalizations are related to the concept of *transcendental numbers* (J. Liouville, Ch. Hermite, F. Lindemann). Nevertheless, we deal only with most elementary geometric ideas.

2. The Steinhaus-Sierpiński Problem. In 1957 Steinhaus raised the following question: *Given the natural number m, does there exist in the plane R^2 of the cartesian coordinates (x_1, x_2) a circle having in its interior exactly m points having integer coordinates?*

Denoting by Z the set of rational integers, it seems appropriate to denote by

$$Z^2 = \{(m_1, m_2)\}$$

the set of points of R^2 having integer coordinates. Z^2 is called the *two-dimensional lattice*. W. Sierpiński (see [**4**, pp. 357–358], [**5**, pp. 8–9] and [**2**, pp. 117–118]) answered Steinhaus's question affirmatively by the following remark: *The point*

$$w = \left(\sqrt{2}, \tfrac{1}{3}\right) \tag{2.1}$$

has different distances from all points of Z^2.

Let us see how this remark solves Steinhaus's problem. We write $b - a$ to denote the vector $\overrightarrow{ab}$ and $|b-a|$ for its length. Clearly Z^2 is a countable set, and so we may write its points as a sequence

$$Z^2 = \{a^1, a^2, a^3, \dots\}.$$

This we do by arranging the points a^ν in the order of *increasing* distances from the point w; hence

$$|a^\nu - w| < |a^{\nu+1} - w|, \qquad (\nu = 1, 2, 3, \dots),$$

which is possible by Sierpiński's remark. It follows that the open circular disk defined by the inequality

$$|x - w| < |a^{m+1} - w|$$

contains the points $a^1, a^2, \dots, a^m$ of Z^2, and no others.

A proof of Sierpiński's remark. Suppose that

$$a = (a_1, a_2), \quad b = (b_1, b_2) \tag{2.2}$$

are two distinct points of Z^2 *at the same distance from the point* w, defined by (2.1), and let us reach a contradiction. Our assumption means that

$$\left(a_1 - \sqrt{2}\right)^2 + \left(a_2 - \tfrac{1}{3}\right)^2 = \left(b_1 - \sqrt{2}\right)^2 + \left(b_2 - \tfrac{1}{3}\right)^2,$$

which amounts to

$$a_1^2 + a_2^2 - b_1^2 - b_2^2 - \tfrac{2}{3}a_2 + \tfrac{2}{3}b_2 = 2(a_1 - b_1)\sqrt{2}\,.$$

The irrationality of $\sqrt{2}$ implies the two relations

$$a_1 - b_1 = 0 \quad \text{and} \quad a_2^2 - b_2^2 - \tfrac{2}{3}(a_2 - b_2) = 0, \tag{2.3}$$

where we already used the first relation to simplify the second. Our assumption (2.2) that $a \neq b$, and $a_1 = b_1$ show that $a_2 - b_2 \neq 0$. We may therefore cancel the factor $a_2 - b_2$ in the second relation (2.3), to conclude that

$$a_2 + b_2 - \tfrac{2}{3} = 0,$$

which is obviously impossible, since a_2, b_2 are integers.

At this point we may ask: How did Sierpiński find his point (2.1)? What is the set of all points having the property of w? These questions seem fair enough; however, we prefer to change the problem by replacing the lattice Z^2 by what we call the *rational lattice*.

3. The Sierpiński Set S_n for the Rational Lattice Q^n. The set of points of R having rational abscissae being denoted as usual by Q, let

$$Q^n = \{(r_1,\ldots,r_n);\quad \text{all}\quad r_i \in Q\} \tag{3.1}$$

be the set of points of R^n having all their coordinates rational. The set Q^n is countable and also dense in R^n; this means that every point of R^n is a limit point of points of Q^n. The main property of Sierpiński's point (2.1) suggests the following:

Problem 1. *To determine the set S_n of points of R^n, such that each point of S_n has different distances from all points of Q^n. In symbols, we can write*

$$S_n = \{x;\, x \in R^n, |x-x'| \neq |x-x''| \quad \text{if} \quad x', x'' \in Q^n, x' \neq x''\}. \tag{3.2}$$

Let us first look at this problem for the simplest case when $n=1$. We are to determine the set S_1 of reals, which are at different distances from all rational points of R.

If x is at *equal* distances from the two *distinct* rationals r_1 and r_2, then

$$x = \frac{r_1 + r_2}{2} \qquad (r_1 \neq r_2), \tag{3.3}$$

and conversely. This point x certainly does not belong to S_1. To obtain the set S_1 we have therefore to *remove* from R all points of the form (3.3). Observe, however, that the point (3.3) is rational; also that every rational r is the average of two distinct rationals, e.g.,

$$r = \frac{(r-1)+(r+1)}{2}.$$

We have just shown that

$$S_1 = R \setminus \bigcup_{r \in Q} r = R \setminus Q \tag{3.4}$$

and established

THEOREM 1. *The set S_1 of points of R, having different distances from all rational points of R, is identical with the set I_1 of irrational points of R.*

This is a (possibly new) definition of irrationals. In some ways it seems to be dual to the usual definition $I_1 = R \setminus Q$ which, I grant you, seems simpler.

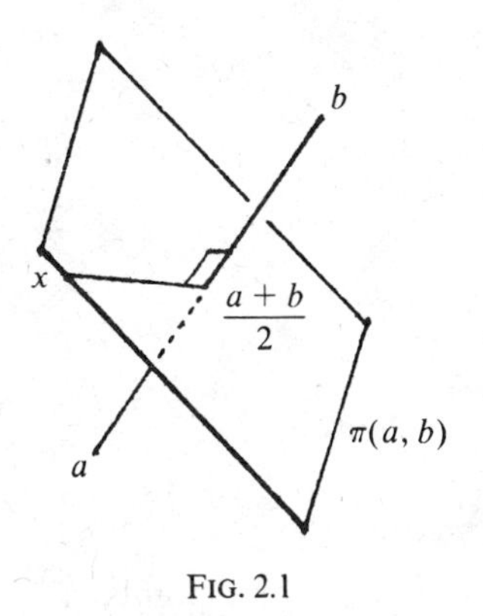

FIG. 2.1

4. Determining the Set S_n. It does seem remarkable that Theorem 1 will generalize to R^n in terms of appropriate definitions. Let us find the locus of points $x=(x_i)$ which are at equal distances from the points a and b of R^n. The distance formula shows the equation of this locus to be

$$\sum_1^n (x_i - a_i)^2 = \sum_1^n (x_i - b_i)^2.$$

Expanding the binomials and canceling Σx_i^2 on both sides, we obtain its equation

$$\sum_1^n \left(x_i - \frac{a_i + b_i}{2} \right)(b_i - a_i) = 0. \tag{4.1}$$

It is seen to pass through the midpoint

$$\frac{a+b}{2} = \left(\frac{a_i + b_i}{2} \right)$$

of the segment $[a, b]$, and to be a plane perpendicular to the vector

$$\overrightarrow{ab} = b - a = (b_i - a_i).$$

We denote it by $\pi(a, b)$ and call it the *perpendicular bisectrix of the points a and b*. Figure 2.1 represents the situation in R^3.

DEFINITION 1. *A plane*

$$\Pi(A_i): A_1 x_1 + \cdots + A_n x_n + A_{n+1} = 0 \qquad \left(\sum_1^n A_i^2 > 0 \right), \tag{4.2}$$

is called a rational plane of R^n, *provided that the coefficients* A_i *are all rationals, or equivalently integers.*

DEFINITION 2. *We say that*

$$x = (x_1, \ldots, x_n) \text{ is a rational point of } R^n$$

provided that it lies on some rational plane of R^n.

For instance $x = (\sqrt{2}, 1)$ is a rational point of R^2 because x lies on the rational line $x_2 - 1 = 0$.

Let us denote by R_n the set of rational points of R^n. We wish to partition R^n into its *rational* and its *irrational points*, and for this reason *we define the set* I_n *of irrational points of* R^n *by*

$$I_n = R^n \setminus R_n; \tag{4.3}$$

hence I_n is the complement of R_n. A more explicit description of I_n is given by

DEFINITION 3. *We say that the point*

$$x = (x_1, \ldots, x_n) \text{ is an irrational point of } R^n$$

provided that a linear relation

$$A_1 x_1 + \cdots + A_n x_n + A_{n+1} = 0,$$

with integer coefficients, implies that all $A_i = 0$.

I hasten to add that this situation is usually described by saying that

the numbers $x_1, x_2, \ldots, x_n, 1$ *are arithmetically linearly independent*. (4.4)

Using the notation of (4.2) we may also express the Definition 2 of the set R_n of rational points of R^n by writing

$$R_n = \bigcup_{A_i \in Z} \Pi(A_i). \tag{4.5}$$

This shows that R_n is the union of all rational planes of R^n.

EXAMPLES. A few illustrations of our definitions are called for.

1. Sierpiński's point $w=(\sqrt{2},\frac{1}{3})$ *is a rational point of* R^2 because it lies on the rational line $3x_2-1=0$.

2. The point

$$x=\left(\frac{\log 2}{\log 7},\frac{\log 3}{\log 7},\frac{\log 5}{\log 7}\right)$$

is an *irrational point of* R^3; hence $x\in I_3$. Indeed, if it were a rational point of R^3, we would have a nontrivial relation $A_1x_1+A_2x_2+A_3x_3+A_4=0$ with integer coefficients; hence

$$A_1\log 2+A_2\log 3+A_3\log 5+A_4\log 7=0.$$

Passing to numbers we obtain

$$2^{A_1}\cdot 3^{A_2}\cdot 5^{A_3}\cdot 7^{A_4}=1,$$

which is impossible by the fundamental theorem of Arithmetic, unless all A_i vanish.

3. One of the landmarks of mathematics was Hermite's theorem (of 1873) that the number e, the base of natural logarithms, is a *transcendental number*. This means that there is no nontrivial equation of the form

$$c_0+c_1e+c_2e^2+\cdots+c_ne^n=0$$

with integer coefficients. By Definition 3, this means that *the point*

$$(e,e^2,\dots,e^n)\in I_n \quad \textit{for all } n.$$

A simple but crucial remark is the set identity

$$\bigcup_{A_i\in Z}\Pi(A_i)=\bigcup_{\substack{a,b\in Q^n\\ a\neq b}}\pi(a,b). \tag{4.6}$$

This will be shown below to hold even elementwise: That every rational plane $\Pi(A_i)$ is also a $\pi(a,b)$ of two distinct points of Q^n. But let us take (4.6) for granted for the moment and use it. Using (4.5) and (4.6), we may rewrite (4.3) as

$$I_n=R^n\setminus\bigcup_{\substack{a,b\in Q^n\\ a\neq b}}\pi(a,b). \tag{4.7}$$

Since the right-hand side describes the set S_n defined by (3.2), we have established

THEOREM 2. *We have the set identity*

$$S_n = I_n, \tag{4.8}$$

i.e., the set S_n of points having different distances from all points of Q^n, is identical with the set I_n of irrational points of R^n (see [3]).

We are still to prove the relation (4.6). 1. That the right-hand side of (4.6) is a subset of its left-hand side is seen from (4.1): If all a_i and b_i are rational, it is clear that (4.1) is a linear equation of the form (4.2) with rational, hence also with integer, coefficients.

2. To show that $\Pi(A_i)$ is identical with some appropriate $\pi(a, b)$, let

$$\Pi(A_i)\colon A_1x_1 + \cdots + A_nx_n + A_{n+1} = 0, A_i \text{ integers} \qquad \left(\sum_1^n A_i^2 > 0\right). \tag{4.9}$$

Let $c = (c_i) \in Q^n$ be any (but fixed) point of this rational plane; hence

$$A_1c_1 + \cdots + A_nc_n + A_{n+1} = 0.$$

Subtracting this from (4.9) we may write

$$\Pi(A_i)\colon \sum_1^n A_i(x_i - c_i) = 0. \tag{4.10}$$

Comparing this with (4.1) we see that we are to determine the a_i and b_i from the equations

$$b_i - a_i = A_i, \quad \frac{a_i + b_i}{2} = c_i \qquad (i = 1, \ldots, n). \tag{4.11}$$

These give the rationals

$$a_i = \tfrac{1}{2}(2c_i - A_i), \quad b_i = \tfrac{1}{2}(2c_i + A_i).$$

Now (4.11) shows that the equation (4.10) reduces to (4.1), hence that

$$\Pi(A_i) = \pi(a, b).$$

The left-hand side of (4.6) is therefore a subset of the right-hand side, proving the identity.

Problems

1. Show that

$$\left(\sqrt{2},\sqrt{3}\right) \in I_2.$$

2. Let $a \in Q^n$, and let Π be a rational plane of R^n. Let b be the mirror image of a with respect to Π. Show that $b \in Q^n$.

3. If $a \in S_n$, then we know that all distances $|a-b|$ $(b \in Q^n)$ are distinct. Show that also *all lines joining the point a to all points b of Q^n are distinct lines.* (Recall that the equations of the straight line joining $a=(a_i)$ to $b=(b_i)$ are

$$\frac{x_1-a_1}{b_1-a_1} = \frac{x_2-a_2}{b_2-a_2} = \cdots = \frac{x_n-a_n}{b_n-a_n}.\Big)$$

Figure 2.2 reminds one of the candelabra at the Metropolitan Opera Company (Lincoln Center, New York City), a gift of the Austrian government. Each candelabrum is a cluster of lucite sticks of different lengths, issuing from the same point a. Our Figure 2.2 covers densely the entire universe.

4. Consider in R^2 the circumference

$$\Gamma: |x-a| = |b-a|$$

having its center at a and passing through b. Show the following:

(i) If $a \in S_2$ and $b \in Q^2$, then

$$\Gamma \cap Q^2 = \{b\}.$$

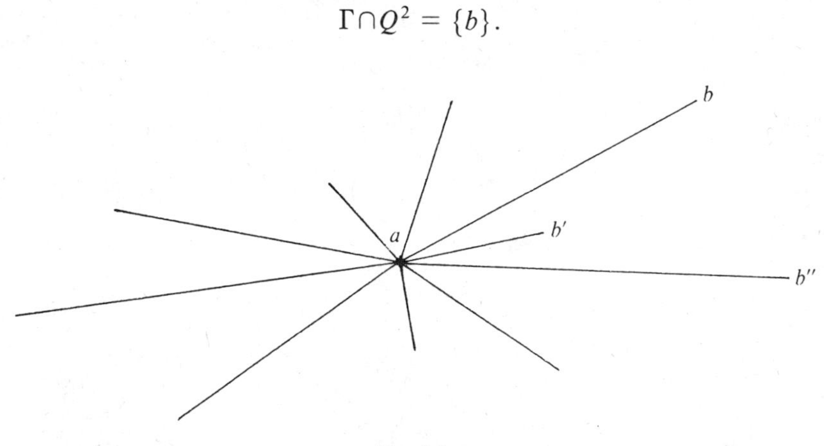

FIG. 2.2

(ii) We now assume that $a \in Q^2$. Show that the intersection

$$\Gamma \cap Q^2$$

is either void or else dense on Γ. (To show (ii) use Problem 2 and consider rational lines through a.)

5. Let τ be a transcendental number, e.g., $\tau = e$, and let

$$t = (\tau, \tau^2, \ldots, \tau^n).$$

In our Examples it was pointed out that $t \in S_n$. Show that

$$\theta = |t - r| \text{ is a transcendental number, where } r \in Q^n.$$

(Show first that θ^2 is transcendental.)

References

[1] G. H. Hardy and E. M. Wright, *An Introduction to the Theory of Numbers*, 3rd ed., Oxford, 1954.

[2] R. Honsberger, *Mathematical Gems*, Dolciani Mathematical Expositions No. 1, Mathematical Association of America, 1973.

[3] I. J. Schoenberg, *On a problem of Steinhaus on lattice points*, Amer. Math. Monthly, 86 (1979) 765–766.

[4] W. Sierpiński, *Elementary Theory of Numbers*, Monografie Matematyczne, Tom 42, Warszawa, 1964.

[5] W. Sierpiński, *A Selection of Problems in the Theory of Numbers*, translated from the Polish by A. Sharma, Macmillan, New York, 1964.

[6] D. J. Struik, *A Concise History of Mathematics*, 3rd ed., Dover, New York, 1967.

CHAPTER 3

A BRIEF INTRODUCTION TO FIBONACCI AND LUCAS NUMBERS

1. Introduction. In the present MTE we use the rudiments of elementary number theory, by which we mean the Euclidean Algorithm and its consequences, in particular the fundamental theorem of arithmetic. We use the symbol $a|b$ for divisibility. We denote by $d=(a, b)$ the g.c.d. of a and b and use the fact that it can be expressed as $d=ax+by$ with integers x and y. In particular, $(a, b)=1$ means that a and b are relatively prime, i.e., have no common divisor exceeding 1. We also use the concept of *congruence modulo* m: The relation $a \equiv b(\text{mod } m)$ means that $m|(a-b)$. Finally, we will rely on what I like to call

EUCLID'S LEMMA. *If*

$$a|bc \quad \textit{and} \quad (a,b)=1, \quad \textit{then} \quad a|c. \tag{1.1}$$

Our main references are the papers **[3]** and **[5]**, but further references will be given as the need arises.

The numbers of the title are two special sequences of integers (x_n) satisfying the recurrence relation

$$x_{n+1} = x_n + x_{n-1} \quad \text{for all integers } n. \tag{1.2}$$

If the values of x_0 and x_1 are prescribed, then (1.2) will define progressively x_n for all n. How do we express x_n explicitly in terms of x_0 and x_1? This is done by a method which applies in the case of any linear recurrence relation. Observe that the sequence $x_n = r^n$ (r fixed) will satisfy (1.2), provided that $r^{n+1}=r^n+r^{n-1}$ for all n, and this requires only that

$$r^2 - r - 1 = 0. \tag{1.3}$$

This quadratic equation has the two roots

$$\alpha = \frac{1+\sqrt{5}}{2}, \quad \beta = \frac{1-\sqrt{5}}{2}. \tag{1.4}$$

By the principle of superposition it follows that

$$x_n = A\alpha^n + B\beta^n \tag{1.5}$$

satisfies (1.2) for any constants A and B. These are determined by requiring that x_0 and x_1 assume the prescribed values.

Among the sequences satisfying (1.2) we single out the two sequences

$$L_n = \left(\frac{1+\sqrt{5}}{2}\right)^n + \left(\frac{1-\sqrt{5}}{2}\right)^n \quad \text{and}$$
$$F_n = \frac{1}{\sqrt{5}}\left\{\left(\frac{1+\sqrt{5}}{2}\right)^n - \left(\frac{1-\sqrt{5}}{2}\right)^n\right\}. \tag{1.6}$$

These are called the *Lucas* and the *Fibonacci numbers*, respectively. Both expressions (1.6) being of the form (1.5), it follows that both sequences (L_n) and (F_n) satisfy (1.2).

But why are they integers? This is true because (1.6) shows that

$$L_0 = 2, \quad L_1 = 1, \tag{1.7}$$

and

$$F_0 = 0, \quad F_1 = 1. \tag{1.8}$$

These initial values x_0 and x_1 being integers, (1.2) takes over and shows that all x_n are integers.

Using their initial values (1.7), (1.8), we get the short table of values

n	0	1	2	3	4	5	6	7	8
L_n	2	1	3	4	7	11	18	29	47
F_n	0	1	1	2	3	5	8	13	21

Even so, it would be irksome to evaluate L_{705}, say.

2. On Lucas Numbers of Prime Subscript. From our table of values we verify that

$$L_3 = 4 \equiv 1 \pmod{3}, \quad L_5 = 11 \equiv 1 \pmod{5}, \quad L_7 = 29 \equiv 1 \pmod{7},$$

where the subscripts 3, 5, 7, are prime numbers. These are not accidents, and we have the following

THEOREM 1. *If p is a prime, then*

$$L_p \equiv 1 \pmod{p}. \tag{2.1}$$

This will follow from Euler's remark of 1736 that

if p is a prime, then the binomial coefficients

$$\binom{p}{1}, \binom{p}{2}, \ldots, \binom{p}{p-1} \text{ are all divisible by } p. \tag{2.2}$$

We first recall the proof of (2.2). In view of

$$\binom{p}{k} = \frac{p(p-1)\cdots(p-k+1)}{k!}, \qquad (k<p),$$

we apply Euclid's Lemma with $a=k!, b=p, c=(p-1)\cdots(p-k+1)$, and find that its assumptions are satisfied. Therefore $a|c$, proving (2.2).

To establish Theorem 1, we use (1.6) with $n=p$ (>2); expanding the binomials, we obtain

$$\begin{aligned} L_p - 1 &= \frac{2}{2^p}\left\{1+\binom{p}{2}5+\binom{p}{4}5^2+\cdots+\binom{p}{p-1}5^{(p-1)/2}\right\}-1 \\ &= \frac{2}{2^p}\left\{\binom{p}{2}5+\cdots+\binom{p}{p-1}5^{(p-1)/2}\right\}-\frac{2^p-2}{2^p}. \end{aligned}$$

Expanding in the numerator the binomial $2^p=(1+1)^p$, we have

$$\begin{aligned} L_p - 1 = {} & \frac{2}{2^p}\left\{\binom{p}{2}5+\cdots+\binom{p}{p-1}5^{(p-1)/2}\right\} \\ & -\frac{1}{2^p}\left\{\binom{p}{1}+\binom{p}{2}+\cdots+\binom{p}{p-1}\right\}. \end{aligned}$$

By (2.2) we may therefore write

$$L_p - 1 = \frac{pc}{2^p} \quad \text{with an integer } c. \tag{2.3}$$

Again the assumptions of Euclid's Lemma are satisfied with $a=2^p, b=p$. Therefore $2^p|c$, and (2.3) implies (2.1).

3. The Fibonacci Pseudoprimes. A frequent source of mathematical discovery is the question whether the converse of a known proposition is also true. In the case of Theorem 1 this question is as follows: *If n is such that the congruence*

$$L_n \equiv 1 \pmod{n} \tag{3.1}$$

holds, is it true that n must be a prime number?

This converse of Theorem 1 is not true. Numbers n which are composite, while (3.1) holds, are known as *Fibonacci pseudoprimes*, which we abbreviate to F.ps. .

Let the sequence

$$N_1, N_2, N_3, \ldots, \tag{3.2}$$

represent all Fibonacci pseudoprimes in increasing order. Very likely the most extensive table of these numbers known at present was constructed in January 1978 by George Logothetis, a graduate student of Computer Science at the University of Wisconsin in Madison. I reproduce it here with his permission.

Table of Fibonacci pseudoprimes below 100 700

k	N_k	k	N_k
1	$705 = 3\cdot5\cdot47$	15	$51\,841 = 47\cdot1103$
2	$2\,465 = 5\cdot17\cdot29$	16	$54\,705 = 3\cdot5\cdot7\cdot521$
3	$2\,737 = 7\cdot17\cdot23$	17	$64\,079 = 139\cdot461$
4	$3\,745 = 5\cdot7\cdot107$	18	$64\,681 = 71\cdot911$
5	$4\,181 = 37\cdot113$	19	$67\,861 = 79\cdot859$
6	$5\,777 = 53\cdot109$	20	$68\,251 = 131\cdot521$
7	$6\,721 = 11\cdot13\cdot47$	21	$75\,077 = 193\cdot389$
8	$10\,877 = 73\cdot149$	22	$80\,189 = 17\cdot53\cdot89$
9	$13\,201 = 43\cdot307$	23	$90\,061 = 113\cdot797$
10	$15\,251 = 101\cdot151$	24	$96\,049 = 139\cdot691$
11	$24\,465 = 3\cdot5\cdot7\cdot233$	25	$97\,921 = 181\cdot541$
12	$29\,281 = 7\cdot47\cdot89$	26	$100\,065 = 3\cdot5\cdot7\cdot953$
13	$34\,561 = 17\cdot19\cdot107$	27	$100\,127 = 223\cdot449$
14	$35\,785 = 5\cdot17\cdot421$		

In the next section we discuss the method by which Logothetis constructed his table. We will also verify here "by hand" that 705 is indeed an F.ps. Since prehistoric times humans have increased the efficiency of their hands by means of tools. One such tool is the hand-held calculator, and we will use it in dealing with the number 705. (See [5].)

Another kind of pseudoprime is defined as follows:

By Fermat's theorem we know that

$$2^n \equiv 2 \pmod{n} \tag{3.3}$$

if n is a prime number. Again the converse proposition is false. Composite numbers n satisfying (3.3) are called *pseudoprimes*, the first being $341 = 11 \cdot 31$. D. H. Lehmer found the *even* pseudoprime

$$161038 = 2 \cdot 73 \cdot 1103.$$

(See Sierpiński's fine book [**8**, Chapter 5, §7].)

Our table of Fibonacci pseudoprimes shows that the first 27 such numbers are all odd. It is not known if there are even Fibonacci pseudoprimes.

4. The Matrix Approach. It consists in replacing the fundamental recurrence relation $x_{n+1} = x_n + x_{n-1}$ by the *vector* recurrence relation

$$\begin{pmatrix} x_n \\ x_{n+1} \end{pmatrix} = \begin{pmatrix} 0 & 1 \\ 1 & 1 \end{pmatrix} \begin{pmatrix} x_{n-1} \\ x_n \end{pmatrix} \tag{4.1}$$

to which it is visibly equivalent. In terms of the 2×2 matrix

$$A = \begin{pmatrix} 0 & 1 \\ 1 & 1 \end{pmatrix}, \tag{4.2}$$

and replacing in (4.1) n successively by $n-1, n-2, \ldots, 1$, we obtain by substitutions that

$$\begin{pmatrix} x_n \\ x_{n+1} \end{pmatrix} = A^n \begin{pmatrix} x_0 \\ x_1 \end{pmatrix}. \tag{4.3}$$

This relation brings to bear on our problem the powerful tool of matrix multiplication.

To verify the congruence

$$L_{705} \equiv 1 \pmod{705}, \tag{4.4}$$

we use (4.3) for $x_n = L_n$; hence $x_0 = 2$, $x_1 = 1$, and $n = 704$, to obtain

$$\begin{pmatrix} L_{704} \\ L_{705} \end{pmatrix} \equiv A^{704} \begin{pmatrix} 2 \\ 1 \end{pmatrix} \quad \pmod{705}. \tag{4.5}$$

This amounts to evaluating the matrix $A^{704} \pmod{705}$. Working modulo 705 means that in performing matrix multiplications, we may discard all multiples of 705.

Even this would be laborious. Fortunately the binary representation

$$704 = 64 + 128 + 512 = 2^6 + 2^7 + 2^9 \tag{4.6}$$

will greatly facilitate our task. By successive squaring of matrices, and working modulo 705 throughout, we start by evaluating the nine matrices

$$A^{2^k} \pmod{705} \quad \text{for} \quad k = 1,2,3,\dots,9. \tag{4.7}$$

In view of (4.6) we select among these the three matrices

$$\begin{aligned} A^{2^6} &\equiv \begin{pmatrix} 142 & 423 \\ 423 & 565 \end{pmatrix}, \quad A^{2^7} \equiv \begin{pmatrix} 283 & 141 \\ 141 & 424 \end{pmatrix}, \\ A^{2^9} &\equiv \begin{pmatrix} 424 & 564 \\ 564 & 283 \end{pmatrix}, \qquad \pmod{705}. \end{aligned} \tag{4.8}$$

Multiplying together these three matrices, mod 705, we see by (4.6) that

$$A^{704} \equiv \begin{pmatrix} 142 & 423 \\ 423 & 565 \end{pmatrix} \pmod{705}. \tag{4.9}$$

Now (4.5) shows that

$$\begin{pmatrix} L_{704} \\ L_{705} \end{pmatrix} \equiv \begin{pmatrix} 142 & 423 \\ 423 & 565 \end{pmatrix} \begin{pmatrix} 2 \\ 1 \end{pmatrix} = \begin{pmatrix} 707 \\ 1411 \end{pmatrix} \equiv \begin{pmatrix} 2 \\ 1 \end{pmatrix} \pmod{705}. \tag{4.10}$$

Therefore the congruence (4.4) indeed holds.

REMARKS. 1. Notice that (4.10) also shows that

$$L_{704} \equiv 2 \pmod{705}. \tag{4.11}$$

Since $L_0 = 2$, $L_1 = 1$, we may summarize both (4.11) and (4.4) by writing that

$$(L_{704}, L_{705}) \equiv (L_0, L_1) \pmod{705}. \tag{4.12}$$

Therefore $L_{706} = L_{705} + L_{704} \equiv L_1 + L_0 = L_2 \pmod{705}$, and likewise $L_{707} = L_{706} + L_{705} \equiv L_2 + L_1 = L_3 \pmod{705}$, and by induction, that

$$L_{n+704} \equiv L_n \pmod{705} \quad \text{for all } n. \tag{4.13}$$

This shows that *the sequence L_n, modulo* 705, *is periodic with the period* 704.

2. If we apply to the second F.ps. 2465 the same method as used above for 705, then instead of (4.10), we obtain

$$\begin{pmatrix} L_{2464} \\ L_{2465} \end{pmatrix} \equiv \begin{pmatrix} 117 & 783 \\ 783 & 900 \end{pmatrix} \begin{pmatrix} 2 \\ 1 \end{pmatrix} = \begin{pmatrix} 1017 \\ 2466 \end{pmatrix} \equiv \begin{pmatrix} 1017 \\ 1 \end{pmatrix} \pmod{2465}. \tag{4.14}$$

This shows that

$$L_{2465} \equiv 1 \pmod{2465}. \tag{4.15}$$

Being a composite number, 2465 is indeed an F.ps. . (4.14) also shows that $L_{2464} \equiv 1017$, hence $L_{2464} \not\equiv 2 \pmod{2465}$, and therefore 2464 is *not* a period of L_n, mod 2465.

3. A few remarks concerning our matrix operations are in order. Observe that A is a *symmetric matrix*, i.e., $A^T = A$. We also know that the product BC of two symmetric matrices that commute ($BC = CB$) is also symmetric. Since any two powers A^m and A^n clearly commute, it follows that all powers A^m are symmetric, as shown by the matrices appearing in (4.8) and (4.9). This means that in multiplying two powers of A we need to compute only one of the two elements off the main diagonal.

The matrix multiplications performed above require the following important check against errors. Passing to determinants, from $|A| = -1$, we conclude that $|A^m| = (-1)^m$. Since all exponents m appearing above are even, we see that $|A^m| = 1$, and, of course, also $|A^m| \equiv 1 \pmod{705}$. *The check is to verify that after each matrix multiplication the resulting product matrix M satisfies the relation $|M| \equiv 1 \pmod{705}$.*

As above we can work out on a hand-held calculator (with a 10-digit display) the matrix $A^{n-1} \pmod{n}$ for any $n < 10^5$. Indeed, all required multiplications, mod n, can be performed exactly because all numbers that we encounter are $< 10^{10}$, the capacity of the calculator.

4. Our last remark concerns the construction of the table of Fibonacci pseudoprimes as given above. Logothetis wrote a program, using the method that gave us the relation (4.10), *to obtain the matrices*

$$A^{n-1} \pmod{n},$$

for all values of $n < 100700$. The final table was obtained by listing only those composite values of n for which

$$L_n \equiv 1 \pmod{n}.$$

The program was an efficient one, as the Computer Center's bill for the computations amounted to \$2.48.

5. Further Applications of the Matrix Approach. Our applications in §4 were mainly computational. We now wish to show how the use of the matrix A allows us to develop *ab initio* some of the best-known properties of the numbers F_n and L_n (see [**1**, §10.14]).

Let us make the relation (4.3), or

$$\begin{pmatrix} x_n \\ x_{n+1} \end{pmatrix} = A^n \begin{pmatrix} x_0 \\ x_1 \end{pmatrix}, \tag{5.1}$$

more explicit by writing

$$A^n = \begin{pmatrix} a_n & b_n \\ c_n & d_n \end{pmatrix}, \tag{5.2}$$

whereby it becomes

$$\begin{aligned} x_n &= a_n x_0 + b_n x_1 \\ x_{n+1} &= c_n x_0 + d_n x_1. \end{aligned} \tag{5.3}$$

This generalizes easily to

$$\begin{aligned} x_{n+k} &= a_n x_k + b_n x_{k+1} \\ x_{n+k+1} &= c_n x_k + d_n x_{k+1}. \end{aligned} \tag{5.4}$$

Indeed, by (5.1),

$$\begin{pmatrix} x_{n+k} \\ x_{n+k+1} \end{pmatrix} = A^{n+k} \begin{pmatrix} x_0 \\ x_1 \end{pmatrix} = A^n \cdot A^k \begin{pmatrix} x_0 \\ x_1 \end{pmatrix} = A^n \begin{pmatrix} x_k \\ x_{k+1} \end{pmatrix},$$

again by (5.1). This and (5.2) show that (5.4) holds.

In (5.3) we let $x_0 = F_0 = 0$ and $x_1 = F_1 = 1$ to obtain that

$$\begin{aligned} F_n &= b_n \\ F_{n+1} &= d_n. \end{aligned} \tag{5.5}$$

Applying (5.4) to $x_n = F_n$ and $k = 1$, observing that $F_1 = 1$, $F_2 = 1$, we obtain

$$\begin{aligned} F_{n+1} &= a_n + b_n \\ F_{n+2} &= c_n + d_n. \end{aligned}$$

These relations and (5.5) show that

$$\begin{aligned} a_n &= F_{n+1} - F_n = F_{n-1}, \\ c_n &= F_{n+2} - F_{n+1} = F_n. \end{aligned}$$

We have therefore shown that

$$A^n = \begin{pmatrix} F_{n-1} & F_n \\ F_n & F_{n+1} \end{pmatrix} \tag{5.6}$$

(see [**2**, Theorem II]). Our previous remark that $|A^n| = (-1)^n$ shows that

$$F_{n+1}F_{n-1} - F_n^2 = (-1)^n, \tag{5.7}$$

a relation derived in the same way as in [**2**, Theorem III]. From (5.6) we see that the elements of the matrices of §4 are appropriate Fibonacci numbers reduced by the moduli 705 and 2465, respectively.

Let us derive the property that

$$F_n \text{ divides } F_{nr}, \text{ where } r \text{ is a natural number.} \tag{5.8}$$

From (5.4) and (5.6) we obtain for $x_n = F_n$ the relation

$$\begin{pmatrix} F_{n+k} \\ F_{n+k+1} \end{pmatrix} = \begin{pmatrix} F_{n-1} & F_n \\ F_n & F_{n+1} \end{pmatrix} \begin{pmatrix} F_k \\ F_{k+1} \end{pmatrix}. \tag{5.9}$$

Replacing here n by nr and k by n, we obtain

$$\begin{pmatrix} F_{n(r+1)} \\ F_{n(r+1)+1} \end{pmatrix} = \begin{pmatrix} F_{nr-1} & F_{nr} \\ F_{nr} & F_{nr+1} \end{pmatrix} \begin{pmatrix} F_n \\ F_{n+1} \end{pmatrix},$$

whence

$$F_{n(r+1)} = F_{nr-1}F_n + F_{nr}F_{n+1}.$$

This shows that, if F_n divides F_{nr}, then F_n also divides $F_{n(r+1)}$, and this proves (5.8) by induction, since (5.8) is obvious if $r = 1$.

As a further application let us establish the property:

$$\text{If } (m, n) = d, \text{ then } (F_m, F_n) = F_d. \tag{5.10}$$

Since d divides m and also n, it follows from (5.8) that

$$F_d \text{ divides } F_m \text{ and also } F_n. \tag{5.11}$$

There remains to show that F_d is the greatest c.d. of F_m and F_n. Let r and s be such that

$$d = mr + ns.$$

From (5.9), on replacing n by mr and k by ns, we obtain

$$\begin{pmatrix} F_{mr+ns} \\ F_{mr+ns+1} \end{pmatrix} = \begin{pmatrix} F_{mr-1} & F_{mr} \\ F_{mr} & F_{mr+1} \end{pmatrix} \begin{pmatrix} F_{ns} \\ F_{ns+1} \end{pmatrix}.$$

This shows in particular that $F_d = F_{mr+ns}$ can be written as

$$F_d = F_{mr-1}F_{ns} + F_{mr}F_{ns+1}. \tag{5.12}$$

By (5.8), *any divisor* δ *of* F_m *and of* F_n also divides F_{mr} and F_{ns} and, by (5.12), that δ *also divides* F_d. Therefore F_d is the greatest c.d. of F_m and F_n, and (5.10) is established.

A last example concerns the Lucas numbers. *Let us show that*

$$L_{n+1}L_{n-1} - L_n^2 = 5\cdot(-1)^{n+1}. \tag{5.13}$$

From (5.1) and (5.6) we have

$$\begin{pmatrix} L_n \\ L_{n+1} \end{pmatrix} = \begin{pmatrix} F_{n-1} & F_n \\ F_n & F_{n+1} \end{pmatrix}\begin{pmatrix} 2 \\ 1 \end{pmatrix}.$$

Again for $x_n = L_n$, but from (5.4) with $k=-1$, we get that

$$\begin{pmatrix} L_{n-1} \\ L_n \end{pmatrix} = \begin{pmatrix} F_{n-1} & F_n \\ F_n & F_{n+1} \end{pmatrix}\begin{pmatrix} -1 \\ 2 \end{pmatrix},$$

because $L_{-1} = -1$, $L_0 = 2$. The last two relations combined give

$$\begin{pmatrix} L_{n-1} & L_n \\ L_n & L_{n+1} \end{pmatrix} = \begin{pmatrix} F_{n-1} & F_n \\ F_n & F_{n+1} \end{pmatrix}\begin{pmatrix} -1 & 2 \\ 2 & 1 \end{pmatrix}.$$

Passing to determinants and using (5.7) we obtain (5.13).

See the paper [**6**] by D. W. Robinson for further arithmetic properties of the Fibonacci numbers. I am indebted for this reference to Olga Taussky Todd.

Problems

1. In the xOy plane divided into unit squares by the lines $x=m$ and $y=n$, we draw the unit square marked by F_1 in Figure 3.1. Next we draw the unit square F_2 to the left of F_1, to be followed by the 2×2 square F_3 below them. Similarly, in spiral fashion we draw the squares $F_4, F_5, F_6, \ldots$. From the relation $F_n = F_{n-1} + F_{n-2}$ we see that the square F_n has its side $= F_n$, the nth Fibonacci number. The diagram amounts to a geometric construction of these numbers.

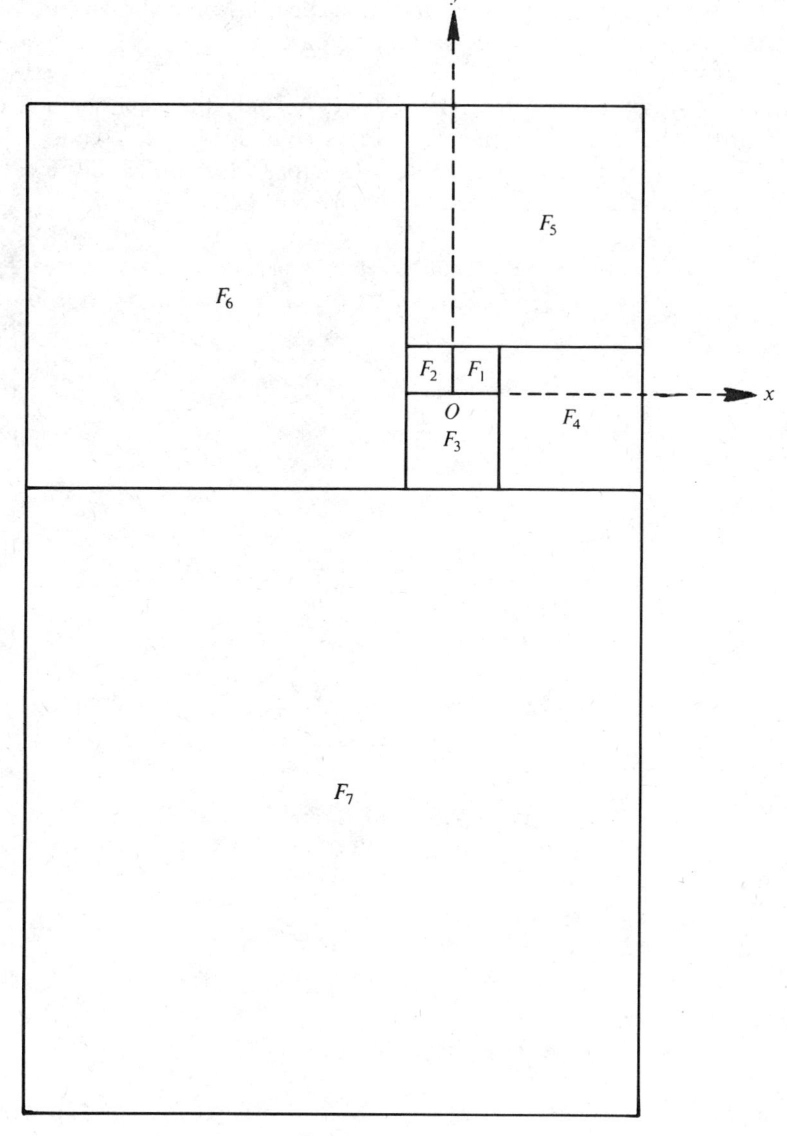

FIG. 3.1

The squares from F_1 to F_n fit together to form a rectangle of dimensions F_n by F_{n+1}. A glance at the diagram shows that we have the relation

$$F_1^2 + F_2^2 + \cdots + F_n^2 = F_n F_{n+1}.$$

Without geometry, establish this relation also by mathematical induction.

2. How many decimal digits do the Lucas numbers L_{705} and L_{100127} have? Use the expression (1.6) of L_n, to show that their numbers of digits are 147 and 162006, respectively. Use logarithms to base 10.

3. In §4 we established the relation (4.10). Apply the same method to derive the relation (4.14) concerning the second Fibonacci pseudoprime 2465. To do this you need a hand-held calculator with a 10-digit display. Be sure to use the check as described in Remark 3 of §4.

4. We say that the natural number A is a period of the sequence L_n, mod a, provided that $L_{n+A} \equiv L_n \pmod{a}$ for all n. Thus in §4 we showed that

$$704 \textit{ is a period of } L_n, \bmod 705. \tag{0}$$

Establish the following lemma:

If

$$k = a_1 a_2 \cdots a_r, \qquad (a_i, a_j) = 1 \quad \text{if} \quad i \neq j, \tag{1}$$

$$A_i \textit{ is a period of } L_n, \qquad \bmod a_i \quad \textit{for all} \quad i, \tag{2}$$

$$A_i | m \quad \textit{for all} \quad i, \tag{3}$$

then

$$m \textit{ is a period of } L_n, \bmod k. \tag{4}$$

This applies nicely for

$$m = 704 \quad \text{and} \quad k = 705 = 3 \cdot 5 \cdot 47 \tag{5}$$

with

$$a_1 = 3, \quad a_2 = 5, \quad a_3 = 47.$$

Indeed, you are asked to check numerically that assumption (2) above holds, the respective periods being

$$A_1 = 8, \quad A_2 = 4, \quad A_3 = 32.$$

Having checked these facts, we observe that also the assumption (3) holds, because 8, 4, and 32 are all divisors of 704. The conclusion (4) establishes in a new way the validity of statement (0) above, or (4.13) of §4. This method of periods was used in [**1**] to show that $L_{705} \equiv 1 \pmod{705}$.

5. Show that the sequence

$$L_n \pmod{m}$$

is always periodic, no matter how the modulus m is chosen.

6. Let (a_n) and (b_n) be the sequences of integers, both satisfying the recurrence relation

$$x_{n+1} = -4x_n - x_{n-1}$$

with the initial values

$$a_0 = 1, \quad a_1 = -2$$

and

$$b_0 = 0, \quad b_1 = 1.$$

Use the methods of the present MTE to establish the following properties:

(i) *If p is a prime, then* $a_p \equiv -2 \pmod{p}$,

(ii) b_n *divides* b_{nr} $(r>1)$,

(iii) $(b_m, b_n) = b_{(m,n)}$.

The sequences (a_n) and (b_n) play a role in equidistant cubic spline interpolation (see [7]).

7. Verify that $n = 341 = 11 \cdot 31$ is a pseudoprime, hence satisfies the congruence (3.3), or

$$2^{341} \equiv 2 \pmod{341}.$$

(Use the hand-held calculator and the binary representation of 341 to evaluate 2^{341} modulo 341.)

References

[1] G. H. Hardy and E. M. Wright, *An Introduction to the Theory of Numbers*, 3rd ed., Oxford, 1954.

[2] V. E. Hoggatt, Jr., *Fibonacci and Lucas Numbers*, Houghton Mifflin, Boston, 1969.

[3] V. E. Hoggatt, Jr., and Marjorie Bicknell, *Some congruences of the Fibonacci numbers modulo a prime p*, Math. Mag., 47 (1974) 210–214.

[4] George Logothetis, Unpublished computations of January 1978 concerning Fibonacci pseudoprimes.

[5] J. M. Pollin and I. J. Schoenberg, *On the matrix approach to Fibonacci numbers and the Fibonacci pseudoprimes*, Fibonacci Quarterly, 18 (1980) 261–268.

[6] D. W. Robinson, *The Fibonacci matrix, modulo m*, Fibonacci Quart., 1 (1963) 29–36.

[7] I. J. Schoenberg, *On equidistant cubic spline interpolation*, Bull. Amer. Math. Soc., 77 (1971) 1039–1044.

[8] W. Sierpiński, *Elementary Theory of Numbers*, Polish Academy of Sciences, Warszawa, 1964.

CHAPTER 4

THE LOCATION OF THE FRETS ON A GUITAR AND A THEOREM OF LORD RAYLEIGH

1. Introduction. The two subjects of the title are not related and are included here because they both originate in the vibrations of strings.

1. The late musicologist J. M. Barbour presented in [**1**] a remarkable discussion of the approximate geometric construction of the location of the frets on a guitar, according to equal temperament, due to the Swedish musicologist D. P. Strähle (1743). The present author was the referee of Barbour's paper and contributed two short footnotes on page 7 of [**1**] which were aimed at simplifying Barbour's analysis. They are the basis of the present account (see [**5**]).

2. A result of Rayleigh [**6**, p. 123] concerns the intertwining of the frequencies of the harmonics of a vibrating string with the similar frequencies of the two strings obtained by dividing the original string into two parts. The motivation for including Rayleigh's result is a problem proposed in the 20th International Mathematical Olympiad in 1979 [**2**].

To introduce our first subject we discuss in §2 the method of tuning keyboard instruments known as *equal temperament* (E.T.). The reader familiar with E.T., or mainly interested in the geometry of our problem, should omit §2 and go directly to §3.

2. The Musical Scale of Equal Temperament. A musical note M has a frequency $(M)=f$, which is the number of complete oscillations per second of the string producing it. If two notes M_1 and M_2 of frequencies f_1 and f_2 are heard simultaneously, then the effect on the ear of the double-note (M_1, M_2) depends mainly on the ratio f_2/f_1. We speak of a *consonant interval*, or *consonant*, provided that this ratio, written in simplest terms, is

$$\frac{f_2}{f_1}=\frac{m_2}{m_1}, \tag{2.1}$$

where m_1 and m_2 are small integers; this was already ancient Greek insight. Disregarding the *unison* when $m_1=m_2=1$, we have the most important case when $m_1=1, m_2=2$, producing the *octave*.

TABLE 1
Consonant Intervals

Interval	Name	Ratio of frequencies
(C, c)	octave	$2/1=2$
(C, G)	fifth	$3/2=1.5$
(C, F)	fourth	$4/3=1.33333$
(C, E)	major third	$5/4=1.25$
$(C, G^{\sharp})$	minor sixth	$8/5=1.6$
$(C, D^{\sharp})$	minor third	$6/5=1.2$
(C, A)	major sixth	$5/3=1.66667$

To describe the remaining consonants we need the notes of the chromatic scale of the octave from C to c, denoted as usual by

$$C, C^{\sharp}, D, D^{\sharp}, E, F, F^{\sharp}, G, G^{\sharp}, A, A^{\sharp}, B, c. \tag{2.2}$$

Table 1 gives the consonant intervals with their names and ratios of frequencies as determined by experiments with monochords. These experiments were possible due to the work of Marin Mersenne (1588–1648), a fellow student and friend of Descartes. He discovered that a uniform taut string of length l has a frequency given by

$$f = \frac{c}{l}, \tag{2.3}$$

where c is a constant depending only on the linear density and tension of the string.

We refer to Rayleigh's treatise [**6**, pp. 8–9], where he derives from Table 1 the frequencies of the notes of the C major scale. Taking the frequency of C as unit, these frequencies, written below the corresponding note, are found to be

$$\frac{C \quad D \quad E \quad F \quad G \quad A \quad B \quad c}{1, \quad \tfrac{9}{8}, \quad \tfrac{5}{4}, \quad \tfrac{4}{3}, \quad \tfrac{3}{2}, \quad \tfrac{5}{3}, \quad \tfrac{15}{8}, \quad 2}. \tag{2.4}$$

This is the so-called *diatonic* scale of C major. This tuning may be continued throughout the entire keyboard by means of accurate octaves. The tuning so obtained seems satisfactory as long as music is performed only in C major, but difficulties arise as soon as the performer tries to *modulate*. As an example, the fifth based on D is the interval (D, A). From (2.4) we find that its ratio of frequencies is

$$\frac{5}{3} \Big/ \frac{9}{8} = \frac{40}{27} = 1.48148.$$

This differs from the theoretical ratio $3/2 = 1.5$ by .01852. This is a relative error of 1.25% which is unacceptable.

The resolution of these difficulties, as early as the 15th century, led to the tuning of keyboard instruments according to the so-called equal temperament (E.T.). *The chromatic scale of* E.T. *is based on the requirement that the ratios of the frequencies between any two consecutive notes of* (2.2) *should be equal.* Since we have in (2.2) twelve such intervals, called semitones, and we wish that $(c)/(C) = 2$, it follows that for each semitone we should have the ratio

$$r = \sqrt[12]{2} = 1.059463094. \tag{2.5}$$

This requirement, of paramount importance, assigns to the notes of (2.2) the frequencies as given in Table 2, where $(C) = 1$. Table 2 shows the frequencies of E.T. The third column repeats the diatonic frequencies of (2.4) and compares them with the frequencies of E.T. The last column lists the relative errors. Notice their small values, which the likes of us would not perceive; Mozart might have been troubled by the errors in A and E. The tuning with the geometric progression of ratio r is continued throughout the keyboard. It is a physical fact of great importance that the natural, or diatonic, scale is so closely approximated by the geometric progression of Table 2.

What makes E.T. *so important*? The answer is that the chromatic scale in (C, c) is musically equivalent to the scale in any other octave. One can play on the instrument in any desired key, and modulation is unlimited. J. S. Bach was a great champion of E.T. To dramatize its advantages he published in 1722 his *Well-tempered Clavier*, containing 24 Preludes and Fugues in all major and minor keys. These could be performed on a keyboard instrument tuned in the equal temperament.

TABLE 2

Note	E.T.	Diatonic scale	Error from E.T.	Relative error
C	$1 = 1$	$C\ 1/1 = 1$	0	0
$C^\sharp$	$r = 1.05946$			
D	$r^2 = 1.12246$	$D\ 9/8 = 1.125$	.00254	.23%
$D^\sharp$	$r^3 = 1.18921$			
E	$r^4 = 1.25992$	$E\ 5/4 = 1.25$	$-.00992$	$-.79\%$
F	$r^5 = 1.33484$	$F = 4/3 = 1.33333$	$-.00151$	$-.11\%$
$F^\sharp$	$r^6 = 1.41421$			
G	$r^7 = 1.49831$	$G\ 3/2 = 1.5$	.00169	.11%
$G^\sharp$	$r^8 = 1.58740$			
A	$r^9 = 1.68179$	$A\ 5/3 = 1.66667$	$-.01512$	$-.91\%$
$A^\sharp$	$r^{10} = 1.78180$			
B	$r^{11} = 1.88775$	$B\ 15/8 = 1.875$	$-.01275$	$-.68\%$
c	$r^{12} = 2$	$c\ 2/1 = 2$	0	0

3. Strähle's Approximate Location of the Frets of a Guitar. We recall Mersenne's result of §2: A uniform taut string of length l has a frequency of vibration given by

$$f = \frac{c}{l}, \tag{3.1}$$

where c is a constant depending only on the linear density and the tension of the string. Let OB be the open string of the guitar producing the note C. Choose OB as the y-axis with origin at O and such that $y=2$ at the point B. Let A be the midpoint where $y=1$.

It follows from (3.1) that the frets, which are to produce the chromatic scale from C to c, are to be placed at the 12 points having the abscissae

$$y = 1, r, r^2, r^3, \ldots, r^{11}, \quad \text{where } r = \sqrt[12]{2}, \tag{3.2}$$

as shown in Figure 4.1. This is a geometric progression of ratio $r = \sqrt[12]{2} = 1.05946$. Equivalently: We are to insert between the point $y=1$ and $y=2$ eleven mean proportionals. As the open string OB sounds the note C, the notes obtained on stopping the string at the frets will be the chromatic scale of Table 2 in §2.

D. P. Strähle's approximate construction shown in Figure 4.2 is as follows. Let VBC be an isosceles triangle having sides of 24, 24, and 12 units. Mark on CV the point A so that $CA=7$ units, and produce BA beyond A so that $OA=AB$. Divide the base CB into 12 equal parts, and join the 12 points of division, including C, to the vertex V. The intersections of the lines so obtained with OB, are the twelve points

$$A, A_1, A_2, \ldots, A_{11}. \tag{3.3}$$

These are the points on OB where the frets are to be placed. The operation of joining a point x, of CB, with V and taking the intersection y, of xV, with OB is called a *perspectivity*, V being the center of perspectivity. We choose CB as x-axis, such that to $x=0$ and $x=1$ correspond the points C and B, respectively. Similarly let OB be the y-axis, such that to $y=0$, $y=1$, $y=2$ correspond the points O, A, B.

We assume as known that the dependence of y on x is of the form

$$y = \frac{\alpha x + \beta}{\gamma x + \delta}. \tag{3.4}$$

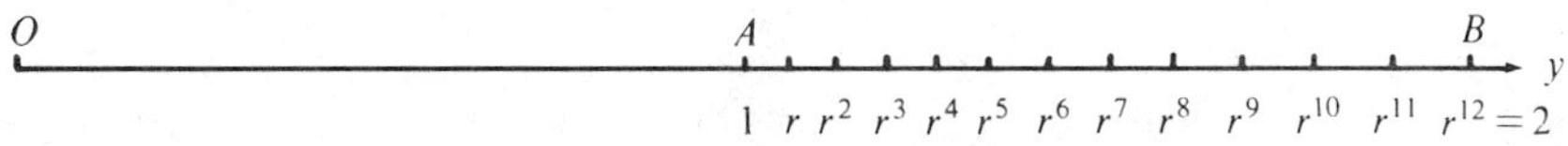

FIG. 4.1

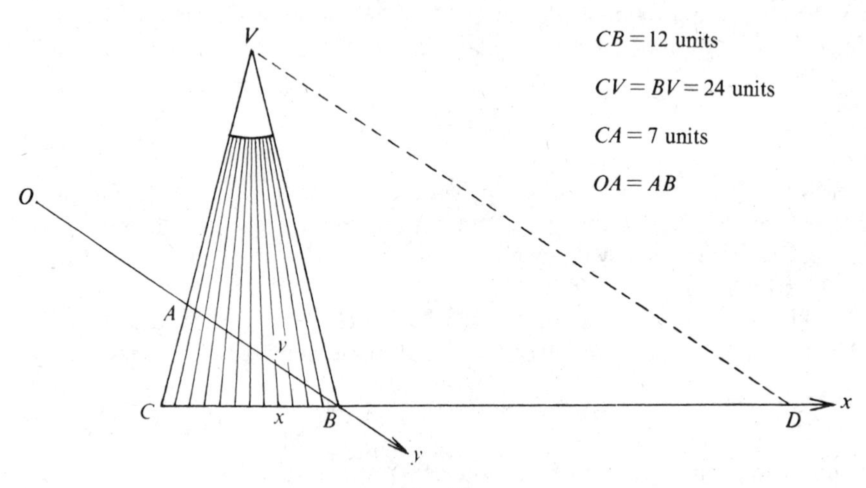

FIG. 4.2

For obvious reasons, this is called a *linear fractional function*, and let us determine its coefficients. We draw VD parallel to OB (Figure 4.2) and intersecting CB in D. By similar triangles we find that $x = CD = 24/7$. However, the perspectivity shows that the corresponding value of y is $y = \infty$, and this shows that (3.4) must be of the form

$$y = \frac{\alpha x + \beta}{\frac{24}{7} - x}.$$

Moreover, Figure 4.2 shows that $x = 0$ and $x = 1$ correspond the values $y = 1$ and $y = 2$, respectively. This gives two linear equations for α and β, having the solutions $\alpha = 10/7$ and $\beta = 24/7$. The final form of the dependence of y on x is therefore

$$y = S(x), \quad \text{where} \quad S(x) = \frac{24 + 10x}{24 - 7x}. \tag{3.5}$$

The y-coordinates of the points (3.3), where the frets are to be placed according to Strähle, are therefore

$$\bar{y}_\nu = S\left(\frac{\nu}{12}\right) \qquad (\nu = 0, 1, \ldots, 11), \tag{3.6}$$

while their locations, according to E.T. as given in Figure 4.1, are

$$y_\nu = 2^{\nu/12} \qquad (\nu = 0, 1, \ldots, 11). \tag{3.7}$$

This raises the question:

$$\textit{How well do the } \bar{y}_\nu \textit{ approximate the } y_\nu? \tag{3.8}$$

An obvious answer would be to compute the values (3.6) and to compare them with the values (3.7) listed in Table 2. It seems more instructive to discuss a preliminary interpolation problem.

4. The Approximation of an Exponential Function by a Linear Fractional Function. Let a and b be given, $0<a<b$, and let us consider the increasing exponential function

$$f(x) = a\left(\frac{b}{a}\right)^x \qquad (0 \leqslant x \leqslant 1). \tag{4.1}$$

Notice that $f(0)=a$, $f(\frac{1}{2})=\sqrt{ab}$, $f(1)=b$. Let us approximate $f(x)$ in $0 \leqslant x \leqslant 1$ by a linear fractional function

$$L(x) = \frac{\alpha x + \beta}{\gamma x + \delta}. \tag{4.2}$$

To obtain it we use interpolation; since $L(x)$ depends on three arbitrary parameters, we choose these such that $L(x)$ will interpolate $f(x)$ at the three points $x=0$, $x=1/2$, and $x=1$. It is easy to verify that

$$L(x) = a\frac{\frac{b}{a}x + \sqrt{\frac{b}{a}}(1-x)}{x + \sqrt{\frac{b}{a}}(1-x)} \tag{4.3}$$

is the required interpolating function because it satisfies the equations

$$L(0) = a = f(0), \quad L(\tfrac{1}{2}) = \sqrt{ab} = f(\tfrac{1}{2}), \quad L(1) = b = f(1). \tag{4.4}$$

If the ratio b/a is not too large, it seems reasonable to expect that $L(x)$, defined by (4.3), will be fairly close to $f(x)$ throughout the interval $0 \leqslant x \leqslant 1$.

Let us look at the special case we are interested in: If $a=1$ and $b=2$, hence

$$f(x) = 2^x, \tag{4.5}$$

then the interpolant (4.3) becomes

$$B(x) = \frac{2x + \sqrt{2}(1-x)}{x + \sqrt{2}(1-x)}, \tag{4.6}$$

where we use the letter B for Barbour, who used this interpolant. For $x=3/4$, where we expect about the largest difference, we find that

$$f\left(\frac{3}{4}\right)=2^{3/4}=1.68179,\quad B\left(\frac{3}{4}\right)=\frac{16-3\sqrt{2}}{7}=1.67962. \tag{4.7}$$

We see that $f(\frac{3}{4})-B(\frac{3}{4})=.00217$. This is a relative error of .13%, which is musically insignificant. We conclude that Barbour's function (4.6) is an acceptable approximation of $f(x)=2^x$.

But how close is Strähle's approximation (3.5)? We know that both $B(x)$ and $S(x)$ interpolate $f(x)=2^x$ at $x=0$ and $x=1$, while they must differ if $x=1/2$. These values are

$$B\left(\frac{1}{2}\right)=\sqrt{2}=\frac{2}{\sqrt{2}}=1.41421 \tag{4.8}$$

and

$$S\left(\frac{1}{2}\right)=\frac{58}{41}=\frac{2}{41/29}=1.41463. \tag{4.9}$$

We see a difference of .00043, which is quite insignificant. Since the days of Christian Huygens (1629–1695) we have known that, in approximating irrational numbers by rational ones, we should use the convergents of their continued fraction expansions. The continued fraction of $\sqrt{2}$ is (see [**4**], Chapter 9, in particular, p. 166)

$$\sqrt{2}=1+\cfrac{1}{2+\cfrac{1}{2+\cfrac{1}{2+\cfrac{1}{2+\cdots}}}}$$

its successive convergents being

$$\frac{1}{1},\frac{3}{2},\frac{7}{5},\frac{17}{12},\frac{41}{29},\ldots. \tag{4.10}$$

Comparing the denominations of the last fractions in (4.8) and (4.9), we see that *we obtain Strähle's approximation from Barbour's by replacing in* (4.8) *the denominator* $\sqrt{2}$ *by the fraction* 41/29, *which is the last convergent* (4.10). By a fundamental theorem of Lagrange (1736–1813) we know that

$$\frac{41}{29}=1.41379$$

is the best approximation of $\sqrt{2} = 1.41421$ *among all rational fractions*

$$\frac{q}{p} \quad \text{with} \quad q \leqslant 41 \quad \text{or} \quad p \leqslant 29.$$

How Strähle discovered his construction remains a mystery, since he had no mathematical training. It must have been by intuition backed by fine craftsmanship. His contribution was not appreciated, because of a numerical mistake made by a contemporary mathematician who was asked to check it and thereby concluded that it was not sufficiently accurate. Barbour discovered this mistake and thereby showed its excellence. See [**1**], also for comparative tables of values and important historical information.

5. A Theorem of Lord Rayleigh on Vibrating Strings. Let AB denote a uniform string under tension of length $= l$. Its frequency f of vibrations ($=$ the number of vibrations per unit of time) is known to be

$$f = \frac{c}{l}, \tag{5.1}$$

where c is a constant depending on the linear density and tension of the string. Assume all units so chosen that $c = 1$ and also that $l = 1$; hence

$$f = 1. \tag{5.2}$$

However, the string AB may also vibrate so that the $n-1$ points, which divide AB into n equal parts, are stationary (Figure 4.3). These points are the *nodes*, and the frequency of vibrations becomes

$$f_n = n, \qquad (n = 1, 2, \ldots). \tag{5.3}$$

The vibration for $n = 1$ is called *fundamental* and those for $n > 1$ are the *harmonics*.

Let me remind the reader that the initial shape problem for the vibrating string AB gave rise to the theory of the Fourier series. A less prestigious but interesting arithmetical application is due to Rayleigh and is as follows.

Let us keep fixed the point C of the string having abscissa $= x$, thereby dividing the string into two parts of lengths x and $1 - x$. By (5.1), with $c = 1$,

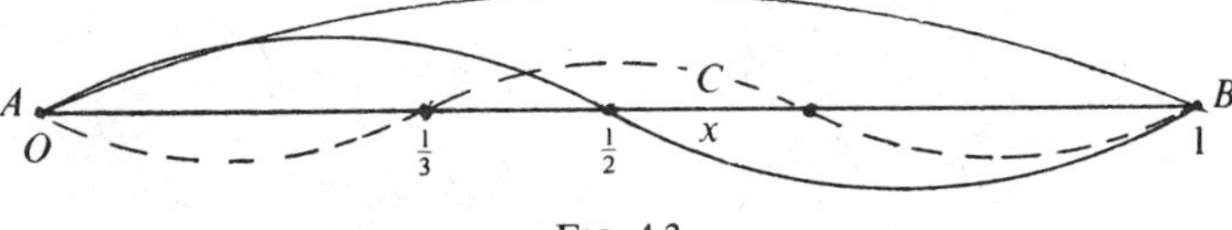

FIG. 4.3

the frequencies of vibrations of the two parts AC and BC are

$$f_n' = \frac{n}{x} \quad \text{and} \quad f_n'' = \frac{n}{1-x} \qquad (0<x<1). \tag{5.4}$$

Rayleigh discovered that between any two consecutive terms of the sequence (5.3) there is one and only one term of the union of the two sequences (5.4) and that this statement becomes particularly neat if we assume x to be irrational. Setting

$$x = \frac{1}{\alpha} \quad \text{and} \quad 1 - x = \frac{1}{\beta},$$

this result assumes the following purely arithmetical form.

THEOREM 1 (Lord Rayleigh). *Let α and β be positive irrational numbers satisfying the equation*

$$\frac{1}{\alpha} + \frac{1}{\beta} = 1, \tag{5.5}$$

and let us define two sequences of integers by

$$A = \{[\alpha n]; n=1,2,\ldots\}, \quad B = \{[\beta n]; n=1,2,\ldots\}, \tag{5.6}$$

where $[t]$ denotes the greatest integer not exceeding the real t; hence $t-1<[t] \leqslant t$. Then

$$A \cap B = \varnothing \tag{5.7}$$

and

$$A \cup B = \{1,2,3,\ldots\}. \tag{5.8}$$

We see that the two sequences are disjoint and fit together to give us all integers. Rayleigh does not bother to give a proof. In §6 we will see that it is quite elementary.

A motivation for including Theorem 1 in this essay is the third problem of the 20th International Mathematical Olympiad of 1979 as presented by S. L. Greitzer in his report **[2]**. Let me call it Problem 1 and reproduce it here.

Problem 1. "*The set of all positive integers is the union of two disjoint subsets $\{f(1), f(2),\ldots,f(n),\ldots\}$, $\{g(1), g(2),\ldots,g(n),\ldots\}$, where*

$$f(1) < f(2) < \cdots < f(n) < \cdots,$$

$$g(1) < g(2) < \cdots < g(n) < \cdots,$$

and $g(n) = f(f(n))+1$ for all $n \geqslant 1$. Determine $f(240)$."

For a student unacquainted with Theorem 1 the problem would seem hopeless. For someone familiar with Theorem 1 it would seem reasonable to assume that

$$f(n) = [\alpha n] \quad \text{and} \quad g(n) = [\beta n], \tag{5.9}$$

for appropriate irrationals satisfying (5.5). If so, the equation

$$g(n) = f(f(n)) + 1 \qquad (n \geq 1) \tag{5.10}$$

amounts to the identity

$$[\beta n] = [\alpha[\alpha n]] + 1. \tag{5.11}$$

As $n \to \infty$ we clearly have

$$\frac{[\beta n]}{n} \to \beta, \quad \frac{[\alpha n]}{n} \to \alpha$$

and therefore

$$\frac{[\alpha[\alpha n]]}{n} = \frac{[\alpha[\alpha n]]}{[\alpha n]} \cdot \frac{[\alpha n]}{n} \to \alpha \cdot \alpha = \alpha^2.$$

Dividing (5.11) by n and letting $n \to \infty$ we therefore obtain the equation

$$\beta = \alpha^2. \tag{5.12}$$

From (5.5) we have $\beta = \alpha/(\alpha - 1)$ and so α must satisfy the equation

$$\alpha^2 - \alpha - 1 = 0, \tag{5.13}$$

whence

$$\alpha = \frac{1+\sqrt{5}}{2} = 1.61803, \quad \beta = \alpha + 1 = 2.61803. \tag{5.14}$$

Therefore $f(240) = [240\alpha] = 388$, which is the required answer.

Have we solved the problem? Hardly so, as we are still to establish

THEOREM 2. *If*

$$\alpha = \frac{1+\sqrt{5}}{2}, \quad \beta = \frac{3+\sqrt{5}}{2}, \tag{5.15}$$

then

$$[\beta n] = [\alpha[\alpha n]] + 1 \quad \textit{for} \quad n = 1,2,\ldots. \tag{5.16}$$

6. Proofs on Theorems 1 and 2. We start with Rayleigh's Theorem 1. This is the kind of result where one can really say that you can't miss the proof. We proceed by contradiction and assume that (5.7) is not true; this means that there are natural numbers m and n such that

$$[\alpha m] = [\beta n]. \tag{6.1}$$

Denoting by q this common value, we have equivalently the inequalities

$$q < \alpha m < q+1, \quad q < \beta n < q+1. \tag{6.2}$$

By solving these for α and β we obtain

$$\frac{m}{q+1} < \frac{1}{\alpha} < \frac{m}{q}, \quad \frac{n}{q+1} < \frac{1}{\beta} < \frac{n}{q}.$$

Adding them together and using (5.5) we have that

$$\frac{m+n}{q+1} < 1 < \frac{m+n}{q}$$

whence

$$q < m+n < q+1.$$

But these inequalities are visibly impossible since we cannot squeeze an integer between two consecutive integers. Hence (6.1) is wrong and (5.7) is established. Notice how essentially the irrationality of α and β was used; without it we could not conclude that (6.2) holds, but only that $q \leqslant \alpha m < q+1$, $q \leqslant \beta n < q+1$.

To prove (5.8) we again assume it to be false and that the union $A \cup B$ leaves gaps in the sequence of unit intervals $(q, q+1)$ $(q = 1,2,\ldots)$. First we observe that either α or β is in the interval $(1,2)$, because if $\alpha > 2$ and $\beta > 2$ then $\alpha^{-1} < 1/2$ and $\beta^{-1} < 1/2$, which would contradict our basic assumption (5.5).

Let us assume, by contradiction, that the interval $(q, q+1)$ $(q \geqslant 2)$ contains no element of $A \cup B$. This assumption is equivalent to the inequalities

$$\alpha m < q < q+1 < \alpha(m+1), \quad \beta n < q < q+1 < \beta(n+1) \tag{6.3}$$

for appropriate integers m and n. Solving these for α and β we obtain that

$$\frac{m}{q} < \frac{1}{\alpha} < \frac{m+1}{q+1}, \quad \frac{n}{q} < \frac{1}{\beta} < \frac{n+1}{q+1}.$$

Adding them and using (5.5) we conclude that

$$\frac{m+n}{q} < 1 < \frac{m+n+2}{q+1},$$

whence

$$m+n < q < q+1 < m+n+2.$$

Again this is impossible since two distinct integers q and $q+1$ have no room between $m+n$ and $m+n+2$. So (6.3) are wrong and (5.8) is established.

Let us now pass to Theorem 2. Since $\beta = \alpha^2$, by (5.12), we are to show that

$$[\alpha[\alpha n]] + 1 = [\alpha^2 n] \quad \text{for all } n. \tag{6.4}$$

This is equivalent to

$$0 < [\alpha^2 n] - \alpha[\alpha n] < 1. \tag{6.5}$$

We prove separately the two inequalities (6.5). From $\alpha^2 = \alpha + 1$ we have $[\alpha^2 n] = [(\alpha+1)n] = [\alpha n + n] = [\alpha n] + n$, and so the left inequality (6.5) amounts to

$$\alpha[\alpha n] < [\alpha n] + n \quad \text{or} \quad (\alpha - 1)[\alpha n] < n.$$

Since $\alpha - 1 = 1/\alpha$, we may write the last inequality as $[\alpha n] < \alpha n$, which is evident.

The second part of (6.5) may be written as

$$[\alpha n + n] < \alpha[\alpha n] + 1 \quad \text{or} \quad [\alpha n] + n < \alpha[\alpha n] + 1.$$

This means that $n - 1 < (\alpha - 1)[\alpha n]$, or

$$\alpha(n-1) < [\alpha n],$$

and this follows from

$$[\alpha(n-1)] < \alpha(n-1) < [\alpha(n-1)] + 1 \leqslant [\alpha n].$$

Notice that the last inequality follows from the property (5.7) of Theorem 1.

7. A Further Structural Property of the Sequences of Theorem 2. The numbers α and β of (5.15) satisfy $\alpha\beta = \alpha + \beta$ and therefore also $\alpha\beta n = \alpha n + \beta n$. This shows that the difference between the integers $[\alpha[\beta n]]$ and $[\alpha n] + [\beta n]$ must be bounded for all n. Actually they are equal as stated in our last theorem. (See page 44.)

THEOREM 3. *If*

$$\alpha = \frac{1+\sqrt{5}}{2}, \quad \beta = \frac{3+\sqrt{5}}{2} \tag{7.1}$$

then

$$[\alpha[\beta n]] = [\alpha n] + [\beta n] \quad \textit{for all } n. \tag{7.2}$$

Proof: From $\beta = \alpha + 1$, hence $n\beta = n\alpha + n$, we conclude that αn and βn have equal fractional parts. This means that

$$\alpha n = [\alpha n] + \theta_n, \quad \beta n = [\beta n] + \theta_n, \qquad (0 < \theta_n < 1), \tag{7.3}$$

and so

$$\alpha n + \beta n = [\alpha n] + [\beta n] + 2\theta_n. \tag{7.4}$$

On the other hand, $\alpha\beta = \alpha + \beta$; hence

$$\alpha\beta n = \alpha n + \beta n. \tag{7.5}$$

Also, by (7.3),

$$\alpha\beta n = \alpha[\beta n] + \alpha\theta_n = [\alpha[\beta n]] + \theta_n' + \alpha\theta_n \qquad (0 < \theta_n' < 1). \tag{7.6}$$

On substituting from (7.3) and (7.6) into (7.5) we obtain that

$$[\alpha[\beta n]] + \theta_n' + \alpha\theta_n = [\alpha n] + [\beta n] + 2\theta_n,$$

or

$$\Delta_n \equiv [\alpha[\beta n]] - [\alpha n] - [\beta n] = (2-\alpha)\theta_n - \theta_n'. \tag{7.7}$$

However, $1 < \alpha < 2$ and so $0 < 2 - \alpha < 1$, and $0 < \theta_n' < 1$. But then (7.7) implies that

$$-1 < \Delta_n < 1 \quad \text{for all } n.$$

Since Δ_n is an integer, we conclude that $\Delta_n = 0$ for all n, and Theorem 3 is established.

Rayleigh's Theorem 1 was rediscovered in 1926 by S. Beàtty of Toronto. For references and a discussion of related problems see Honsberger's book [**3**, Essay Twelve, 93–106].

Problems

1. Let

$$B(x) = \frac{2x+\sqrt{2}(1-x)}{x+\sqrt{2}(1-x)}$$

be the linear fractional function (4.6) interpolating $f(x)=2^x$ at $x=0,1/2,1$. Define a new function $s(x)$ for all real x as follows:

(i) Let

$$s(x) = B(x) \quad \text{if} \quad 0 \leqslant x \leqslant 1.$$

(ii) Extend the definition of $s(x)$ to all real x by means of the functional equation

$$s(x+1) = 2s(x) \qquad (-\infty<x<\infty).$$

Show

(a) that $s(x)$ interpolates 2^x at all integers n and also at $n+\frac{1}{2}$,

(b) that $s(x)$ has a continuous derivative $s'(x)$ for all real x. The function $s(x)$ may be called a *linear fractional spline.*

Hint: Verify the relation $B'(1)=2B'(0)$, which implies the continuity of $s'(x)$.

2. Let α and β be positive numbers such that the equations

$$[\beta n] = [\alpha[\alpha n]]+1 \quad [\alpha[\beta n]] = [\alpha n]+[\beta n]$$

are satisfied for all $n=1,2,3,\ldots$. Show that

$$\alpha = \frac{1+\sqrt{5}}{2}, \quad \beta = \frac{3+\sqrt{5}}{2}.$$

3. Out of context, we propose here the second problem of the 20th International Mathematical Olympiad (see [2]) because of its beauty:

"P is a given point inside a given sphere and A, B, C, are any three points on the sphere such that PA, PB, and PC are mutually perpendicular. Let Q be the vertex diagonally opposite to P in the parallelepiped determined by PA, PB, and PC. Find the locus of Q."

Hint: First solve the problem in the plane: P is given inside a given circle, and A and B are on the circle such that $PA \perp PB$. If Q is the 4th vertex of the rectangle $APBQ$, find the locus of Q. Solve this problem and find a similar solution for the Olympiad problem. Professor Greitzer informed me that Murray Klamkin is the author of the second Olympiad problem.

References

[1] J. M. Barbour, *A geometrical approximation to the roots of numbers*, Amer. Math. Monthly, 64 (1957) 1–9.

[2] S. L. Greitzer, *The Twentieth International Mathematical Olympiad*, Amer. Math. Monthly, 86 (1979) 747–749.

[3] Ross Honsberger, *Ingenuity in Mathematics*, Mathematical Association of America, 1970.

[4] W. J. LeVeque, *Topics in Number Theory*, vol. 1, Addison-Wesley, Reading, Mass., 1956.

[5] I. J. Schoenberg, *On the location of the frets on a guitar*, Amer. Math. Monthly, 83 (1976) 550–552.

[6] Lord Rayleigh, *The Theory of Sound*, vol. 1, 2nd ed., Dover Publications, New York, 1945.

CHAPTER 5

HELLY'S THEOREM ON CONVEX SETS

1. Introduction. Ancient Greek plane geometry studied such simple figures as points, straight lines, polygons, circles, ellipses, and a few further curves defined as loci of a point with appropriate properties. A closer study of functions during the 19th century showed the need to investigate more general sets of points. This gave rise to the theory of sets pioneered by G. Cantor (1845-1918). A classification of sets led to open sets, closed sets, as well as to such operations on sets A and B as the union of sets, $A \cup B$, and their intersection $A \cap B$. We assume familiarity with such basic notions.

Even before the advent of set theory, the so-called *convex sets* attracted the attention of geometers due to their simple and striking properties. A set C is said to be *convex*, provided that, if P and Q are points of C, then the entire segment $[P, Q]$ belongs to C. Not only are figures bounded by triangles, circles, and ellipses convex, but the intersection of any number of convex sets is either void or convex. Notice that a circular disk remains convex even if we remove from its circumference any set of points.

A remarkable result on convex sets discovered as late as 1923 by E. Helly (1884–1943) [**1**] is the following:

THEOREM 1 (Helly). *If $C_1, C_2, \ldots, C_m$ is a finite collection of convex sets in the plane ($m \geq 3$), such that every three among them have a common point, i.e.,*

$$C_i \cap C_j \cap C_k \neq \varnothing \quad \text{if} \quad i < j < k, \tag{1.1}$$

then all m sets have a common point, i.e.,

$$\bigcap_{i=1}^{m} C_i \neq \varnothing. \tag{1.2}$$

Notice that the theorem is no longer valid if we only require that any *two* of the sets should have a common point; the three sides of a triangle furnish a counterexample.

Here is a second related result. In 1949 I found in A. Tarski's preprint [**3**] the following theorem, which he attributes to M. Dresher and T. E. Harris (unpublished).

THEOREM 2 (Dresher-Harris). *Let $S_1, S_2, \ldots, S_m$ be m parallel line segments in the plane, no two of which lie on the same line. If every three among them*

$$S_i, S_j, S_k \qquad (i<j<k)$$

can be intersected by an appropriate straight line, then there exists a straight line intersecting all m segments.

An obviously equivalent "traffic formulation" of Theorem 2 is this: On a straight road Ox we have m traffic signals $T_1, T_2, \ldots, T_m$, such that T_ν shows "green" only during the finite time interval

$$a_\nu \leqslant t \leqslant b_\nu \qquad (\nu = 1, \ldots, m),$$

and "red" otherwise. If these time intervals are such that an appropriate uniformly driving car can pass through every triple of lights

$$T_i, T_j, T_k \qquad (i<j<k), \tag{1.3}$$

then there is a uniformly driving car that can pass through all m lights.

Figure 5.1 shows in the xOt plane the motion of a car passing the traffic lights (1.3).

In §§2 and 3 we offer a proof of Helly's Theorem 1 following the approach of [**2**]. In §4 it is shown how Theorem 1 easily implies Theorem 2. Can Theorem 2 be established directly without applying Helly's Theorem? The answer is yes, and the reader is invited to find a direct proof by induction in m, or otherwise. Just such a direct proof of Theorem 2, due to Erwin Kleinfeld, is the subject of §5.

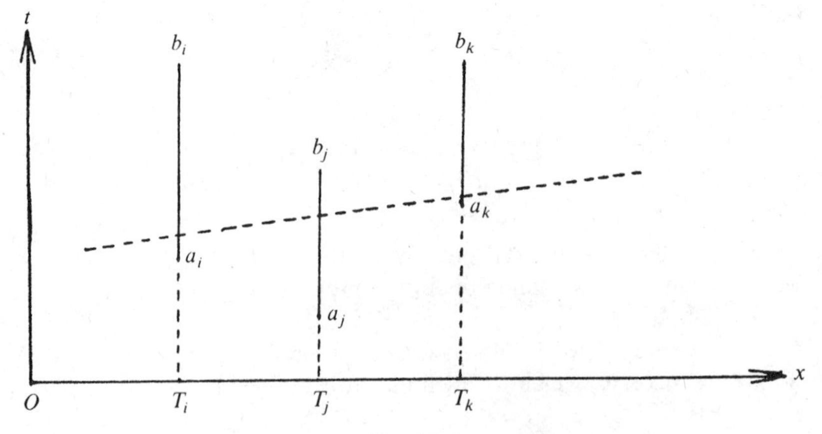

FIG. 5.1

2. A Proof of Theorem 1 When the C_i Are Closed and Bounded Sets. We make the following two assumptions:

$$\textit{If} \quad i<j<k, \quad \textit{then} \quad C_i \cap C_j \cap C_k \neq \varnothing, \tag{2.1}$$

and

$$\bigcap_{1}^{m} C_i = \varnothing, \tag{2.2}$$

and wish to reach a *contradiction*. Notice that (2.1) is the assumption of Theorem 1 and that (2.2) is the negation of its conclusion (1.2).

We say that the circular disk

$$\Delta = \Delta(P,r) = \{Q; PQ \leqslant r\}, \tag{2.3}$$

having center P and radius r, is a *contact disk* of the sets $C_1,\ldots,C_m$, provided that it makes contact with each of the sets, i.e.,

$$\Delta \cap C_i \neq \varnothing \qquad (i=1,\ldots,m). \tag{2.4}$$

Among all such contact disks Δ we look for a disk having the least radius.

Can you show the existence of such a contact disk of least radius? This requires the usual arguments using minimizing sequences, the Bolzano-Weierstrass theorem, and the compactness of the C_i. In fact we made the assumption of the boundedness and closure of the C_i so that these arguments should work nicely. Here we take the existence for granted and wish to find properties of the smallest Δ.

In Figure 5.2 we show this least contact disk $\Delta = \Delta(P,r)$ and let us derive some of its necessary properties. In the first place, as it makes contact with all C_i, it is clear that the sets

$$\Delta \cap C_i \qquad (i=1,\ldots,n) \tag{2.5}$$

are all nonvoid and convex. The sets (2.5) are of two kinds, which we call (a) and (b), depending on whether they reduce to a single point or not. Figure 5.2 shows that $C_1 \in$ (a) and $C_m \in$ (b). Since our Δ is a smallest contact disk, there must be some cases (a), or else, keeping P fixed, we could diminish r a little and still have a contact disk.

Let all cases (a) occur for the sets

$$C_1, C_2, \ldots, C_k \qquad (k \geqslant 1), \tag{2.6}$$

and all cases (b) for the remaining sets

$$C_{k+1}, C_{k+2}, \ldots, C_m. \tag{2.7}$$

This is no restriction, being a matter of appropriately labeling the sets. Of course, we could have $k=m$, when cases (b) do not occur. Let

$$P_i = \Delta \cap C_i \qquad (i=1,\ldots,k) \tag{2.8}$$

be the sets (2.5) which reduce to single points, which are necessarily on the boundary of Δ. Again without restricting the situation we may assume that the k points (2.8) follow in order along the circumference of Δ so that the polygon

$$K = P_1 P_2 \ldots P_k \tag{2.9}$$

is a simple closed and *convex* polygon.

Notice, however, that in Figure 5.2 the center P of Δ is *outside* of the polygon K. We drew the figure on purpose this way, because we want to show that *this situation cannot occur.* In other words: *The contact disk* $\Delta(P, r)$ *shown in* Figure 5.2 *is not a* smallest *contact disk.*

Proof: Let PN be the perpendicular from P to the nearest side of K, which happens to be P_1P_k. If we pick a point P' on PN, between P and N,

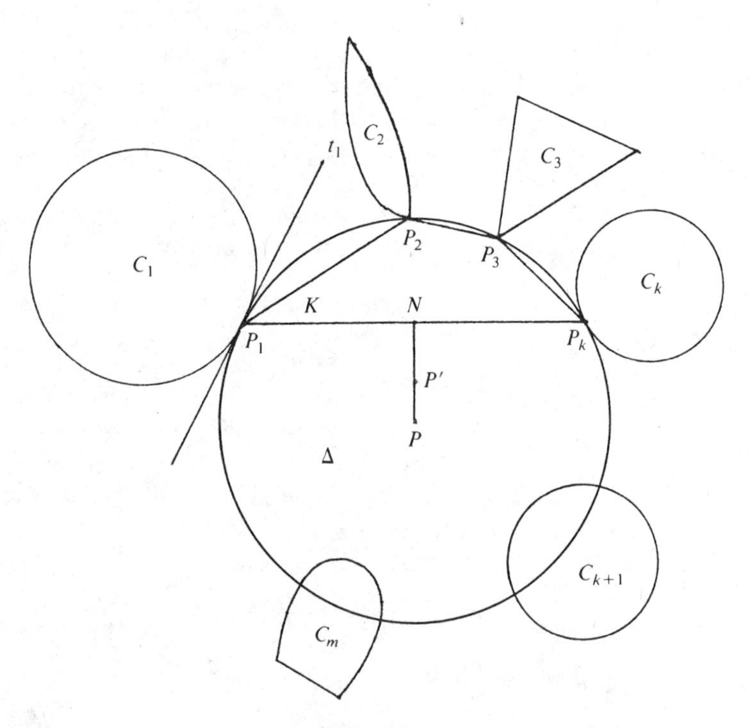

FIG. 5.2

then P' will be nearer than P to any point on P_1P_k, or above P_1P_k. In particular we have that

$$P'P_i < PP_i = r \quad \text{for} \quad i = 1,\ldots,k. \tag{2.10}$$

Let

$$r' = \max P'P_i \qquad (i=1,\ldots,k) \tag{2.11}$$

and let us look at the disk

$$\Delta' = \Delta(P', r'). \tag{2.12}$$

By (2.7) and (2.8) we have $r' < r$ so that Δ' is smaller than Δ. Moreover, Δ' surely makes contact with the sets (2.6). However, it will certainly make contact also with each of the sets (2.7), provided that P' is sufficiently close to P.

Our conclusion: *If $\Delta(P, r)$ is a smallest contact disk, then its center P must belong to the convex polygon* (2.9); hence

$$P \in K. \tag{2.13}$$

Since $r > 0$, (2.13) shows that we must have that

$$k \geq 2. \tag{2.14}$$

The inclusion (2.13) can occur in two different ways depending on whether P belongs to the boundary of K or to the interior of K.

Case A. *The center P of Δ is on a side P_1P_k of K*, which is therefore a diameter of $\Delta(P, r)$. This case is shown in Figure 5.3(a).

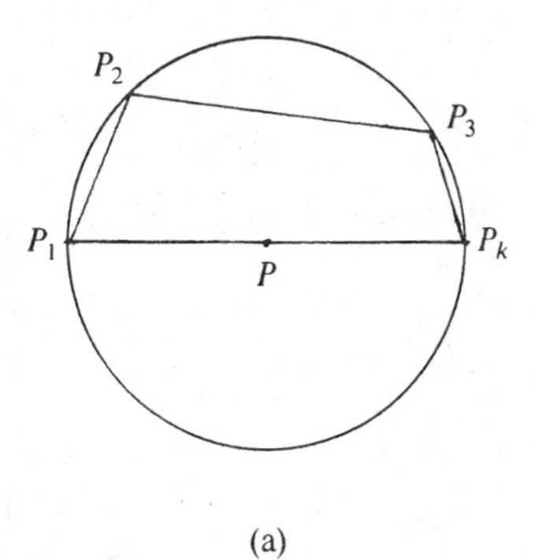

(a)

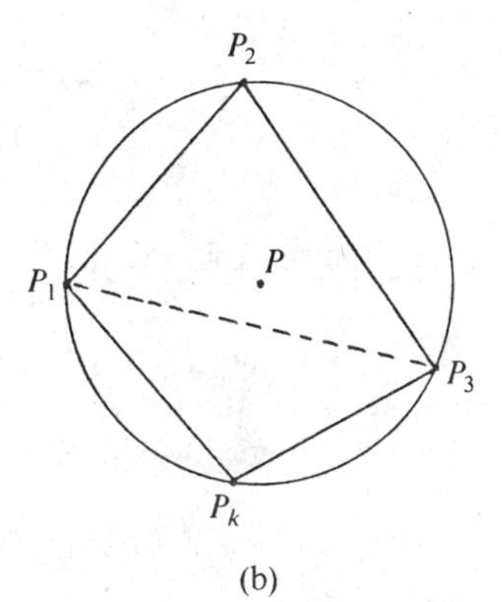

(b)

FIG. 5.3

Case B. *The center P is in the interior of $K = P_1P_2\ldots P_k$*. We triangulate the polygon K. If P is on a side of one of the triangles, then this side is a diameter. If not, then P is interior to one of the triangles.

We have reached one of the following two cases:

Case (a): Two of the points $P_1, P_2, \ldots, P_k$ are the endpoints of a diameter of $\Delta(P, r)$, Figure 5.3(a).

Case (b): Three of the points $P_1, \ldots, P_k$, say P_1, P_2, P_3, are the vertices of an acute-angled triangle containing the center P in its interior, Figure 5.3 (b).

We claim that *either of these two possible conclusions leads to a contradiction with our basic assumption* (2.1).

Proof: Referring to Figure 5.4, let t_i denote the tangent to the disk at the point P_i. Also let H_i denote the closed half-plane, bounded by t_i, which contains the set C_i:

$$C_i \subset H_i. \tag{2.15}$$

Observe next that

$$H_1 \cap H_2 = \varnothing \quad \text{in the Case (a)}, \tag{2.16}$$

and

$$H_1 \cap H_2 \cap H_3 = \varnothing \quad \text{in the Case (b)}. \tag{2.17}$$

However, (2.16) and (2.15) imply that

$$C_1 \cap C_2 = \varnothing. \tag{2.18}$$

Likewise (2.17) and (2.15) imply that

$$C_1 \cap C_2 \cap C_3 = \varnothing. \tag{2.19}$$

Clearly, either of these two conclusions contradicts our assumption (2.1).

3. A Proof of Theorem 1 for Convex Sets. We reduce this case to the previous one already settled. By a combinatorial argument we construct below bounded and closed convex sets D_i such that

$$D_i \subset C_i \qquad (i = 1, \ldots, m), \tag{3.1}$$

$$\text{if} \quad i < j < k, \quad \text{then} \quad D_i \cap D_j \cap D_k \neq \varnothing. \tag{3.2}$$

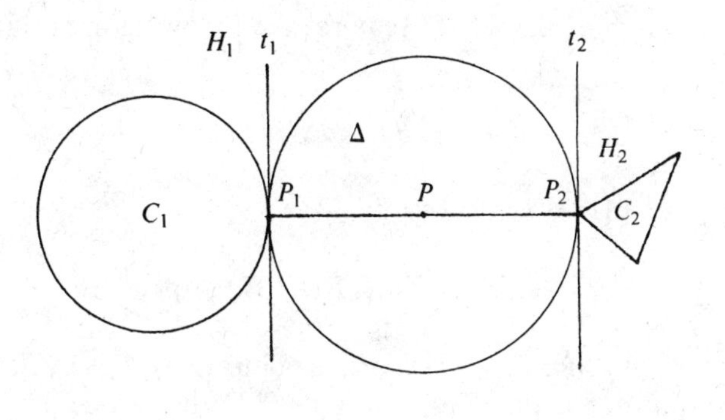

(a)

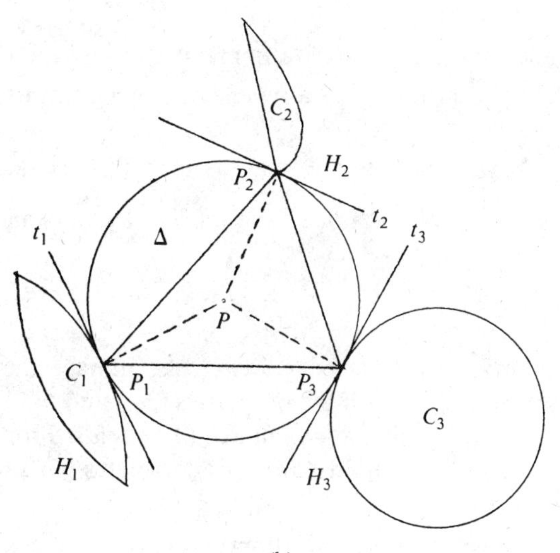

(b)

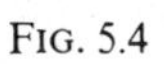

Fig. 5.4

By the previous case we know that

$$\bigcap_1^m D_i \neq \varnothing\,, \tag{3.3}$$

and therefore, by (3.1), also

$$\bigcap_1^m C_i \neq \varnothing\,. \tag{3.4}$$

The construction of the sets D_i requires the following important concept. If

$$A, B, \ldots, F, G, \tag{3.5}$$

are points in the plane, we denote, by

$$K = K(A, B, \ldots, G), \tag{3.6}$$

the *convex hull*, or *convex extension* of the points (3.5). K is by definition the least convex polygon containing the points (3.5); its vertices are among the points (3.5), but not necessarily all of them, as some may be inside K.

For every triple (i, j, k), with $i<j<k$, we pick a point $A_{i,j,k}$ such that

$$A_{i,j,k} \in C_i \cap C_j \cap C_k. \tag{3.7}$$

By (1.1) we know that there is such a point. We obtain thereby $\binom{m}{3}$ such points $A_{i,j,k}$; let us regard $A_{i,j,k}$ as a *symmetric* function of its subscripts, so that $A_{i,j,k} = A_{j,k,i} = \cdots$.

If we keep i fixed, then j and k may assume $\binom{m-1}{2}$ distinct pairs of values; we form the convex hull of all these $\binom{m-1}{2}$ points and denote it by

$$D_i = K\{A_{i,j,k};\, i \quad \text{fixed}, \quad j<k, j \neq i, k \neq i\}. \tag{3.8}$$

These are m bounded and closed convex sets, in fact they are convex polygons. These are the sets with the properties (3.1) and (3.2). Indeed, (3.7) shows that $A_{i,j,k} \subset C_i$, for all $j, k \neq i$, and so (3.1) holds. From $A_{i,j,k} \in D_i \cap D_j \cap D_k$ we see that also (3.2) holds, and this completes our proof.

4. A Proof of Theorem 2 Follows from Theorem 1. We choose an xOy coordinate system and assume the m segments S_i to be defined by

$$S_i = \{(x, y);\quad x = x_i, a_i \leq y \leq b_i\} \qquad (x_1 < x_2 < \cdots < x_m). \tag{4.1}$$

A line

$$l\colon y = \xi x + \eta \tag{4.2}$$

intersects S_i, provided that

$$a_i \leq \xi x_i + \eta \leq b_i \tag{4.3}$$

(see Figure 5.5(a)). These inequalities we now interpret in the (ξ, η)-plane to

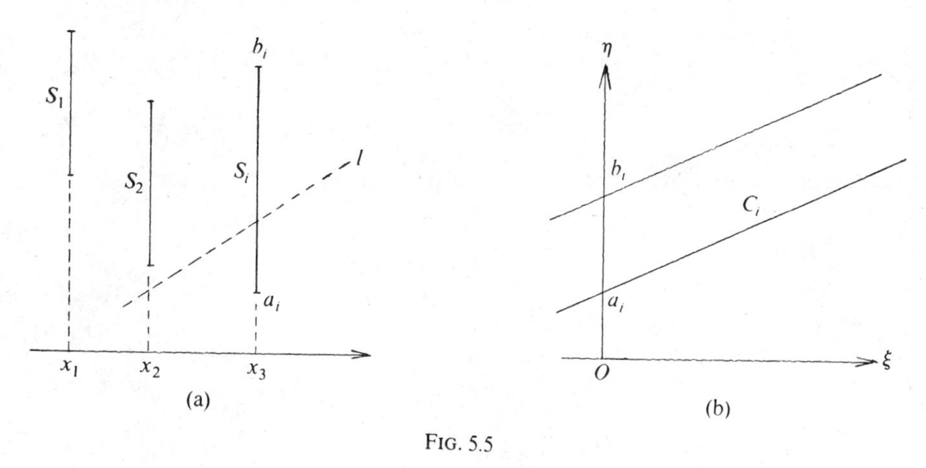

FIG. 5.5

mean that the point (ξ, η) is in the infinite parallel strip

$$C_i = \{(\xi, \eta); a_i \leqslant \xi x_i + \eta \leqslant b_i\} \tag{4.4}$$

(see Figure 5.5(b)), bounded by the two parallel lines

$$\eta = -\xi x_i + a_i, \quad \eta = -\xi x_i + b_i.$$

The assumption of Theorem 2 that some line (4.2) should intersect the segments S_i, S_j, S_k, means in our interpretation that

$$C_k \cap C_j \cap C_k \neq \varnothing \quad \text{if} \quad i < j < k.$$

As the strips (4.4) are convex sets, by Helly's Theorem 1, there is a point (ξ, η) such that

$$(\xi, \eta) \in \bigcap_1^m C_i.$$

The corresponding line (4.2) intersects all m segments (4.1).

5. A Direct Proof of the Dresher-Harris Theorem 2 due to Erwin Kleinfeld. Having proposed a proof of Theorem 2 as a problem in the Rademacher-Schoenberg Problem Seminar at the University of Pennsylvania in 1949, Erwin Kleinfeld, a member of the seminar, now at the University of Iowa, gave the following elegant solution.

As the theorem is trivial if $m = 3$, we assume that $m > 3$ and apply mathematical induction in m. Let the m segments be

$$S_i = [A_i, B_i], \quad A_i = (x_i, a_i), \quad B_i = (x_i, b_i), \quad a_i < b_i,$$
$$x_1 < x_2 < \cdots < x_m. \tag{5.1}$$

We set aside S_m for the moment, and the induction assumption tells us that

$$S_1, S_2, \ldots, S_{m-1} \tag{5.2}$$

are intersected by at least one straight line. Among these intersecting lines *let*

$$L = L'L \textit{ be the line of least slope}, \tag{5.3}$$

and let

$$M = M'M \textit{ be the line of largest slope}, \tag{5.4}$$

both L and M being shown in Figure 5.6. We assume that these lines intersect, and let $V = L \cap M$ be their intersection. Call Ω the double-angle

$$\Omega = (\angle LVM) \cup (\angle L'VM'),$$

whose points are between L and M. Clearly, each of the segments (5.2) intersects both lines L and M.

A proof will follow from several statements, the first of which is as follows:

$$\textit{The line } L \textit{ must contain a pair of endpoints } B_\alpha \textit{ and } A_\beta, \textit{ with } \alpha < \beta. \tag{5.5}$$

Indeed, let $B_{\alpha_1}, B_{\alpha_2}, \ldots,$ be all the upper endpoints on L and $A_{\beta_1}, A_{\beta_2}, \ldots,$ be all the lower endpoints on L. If we had

$$\beta_r < \alpha_s$$

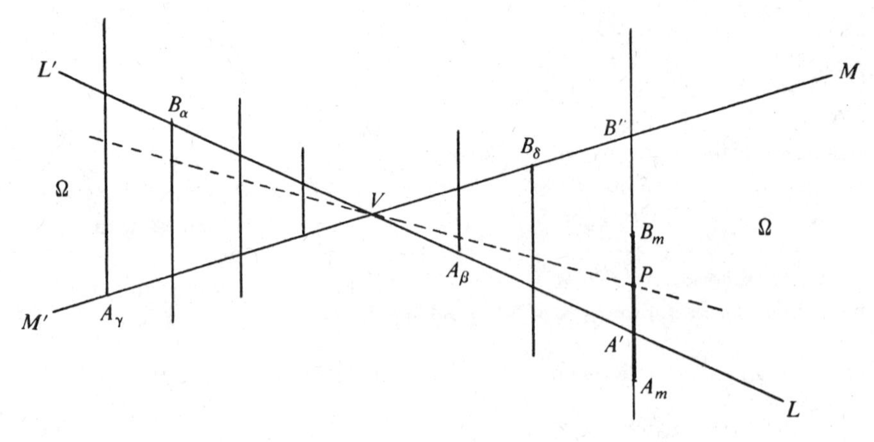

FIG. 5.6

for all these endpoints, then nothing would prevent us from turning L slightly clockwise, thereby further diminishing its slope, while preserving its property of intersecting all segments (5.1). This would contradict the definition (5.3) of L.

The line M must contain a pair of endpoints A_γ and B_δ, with $\gamma < \delta$. (5.6)

A proof similar to the one for (5.5) shows that if (5.6) were not true we could still increase the slope of M.

We now bring in the last segment S_m and observe that we must have that

$$S_m \cap \Omega \neq \varnothing . \tag{5.7}$$

This we derive from (5.1), (5.5), and (5.6) as follows. Let the line $x = x_m$ intersect L and M in A' and B', respectively (Figure 5.6). If (5.7) does not hold, hence

$$S_m \cap \Omega = \varnothing ,$$

then we must have one of the following two situations:
Either

$$B_m \text{ is below } A' \tag{5.8}$$

or else

$$A_m \text{ is above } B'. \tag{5.9}$$

However, if (5.8) holds then Figure 5.6 shows that the three segments

$$S_\alpha, \quad S_\beta, \quad \text{and} \quad S_m$$

could not possibly be intersected by any line, contradicting the assumption that any three among the segments (5.1) can be so intersected. Again, if (5.9) holds, then the segments

$$S_\gamma, \quad S_\delta, \quad \text{and} \quad S_m$$

cannot be intersected by a line, leading to the same contradiction.

We have just shown that (5.7) holds. If

$$P \in S_m \cap \Omega, \tag{5.10}$$

then the line PV, joining P to V, is entirely contained in Ω and must therefore intersect all segments (5.1). *The reason*: Each segment (5.2) intersects both L and M, and therefore the line VP intersects every segment (5.2). Now (5.10) implies that $P \in S_m$, and so VP intersects all segments (5.1).

Let me close with a little story showing the change in manners and morals during the last 30 years. I spoke about our subject at a prestigious college for women in the spring of 1949, announcing the title of the present essay. The head of the department wrote to me: "... If possible, we should like an alternative title which would omit the name of the author. Can you imagine what an irreverant student-body who have never heard of the mathematician might do to the name of Helly?" I complied and spoke about "A theorem on convex domains with applications." This would be unthinkable nowadays.

Problems

1. Let m points be given in the plane such that each three of them can be enclosed in a circle of radius $= r$. Prove that all m points can be enclosed in a circle of radius r.

2. Let $S = \{P_1, \ldots, P_m\}$ be a set of m points in the plane having the diameter $d = \max P_iP_j$. Prove that all m points can be covered by a circular disk of radius $r = d/\sqrt{3}$.

Hint: Notice that a regular triangle having all its sides $= d$ has a circumscribed circle of radius $= d/\sqrt{3}$. Apply the result of Problem 1.

3. Helly's Theorem 1 generalizes to the n-dimensional space R^n. The definition of a convex set is the same as for the plane. The theorem states that, if $C_1, \ldots, C_m$, $(m \geqslant n+1)$ is a finite collection of convex sets in R^n, such that every $n+1$ among them have a common point, then all m sets have a common point.

Analyze our proof as given in §§2 and 3, and find a proof of the theorem for R^3 $(n=3)$. The proof is again by contradiction. The disk $\Delta(P, r)$ of §2 has to be replaced by a solid ball $S = S(P, r)$, and the proof is based on the necessary properties of a *contact ball having the least radius*. The two possibilities of Fig. 5.3 are now to be replaced by three: 1. The two points P_1 and P_2 on the surface of S are the endpoints of a diameter of S. 2. The three points P_1, P_2, P_3 are vertices of an acute-angled triangle inscribed in a great circle of S. 3. The four points P_1, P_2, P_3, P_4 are the vertices of a tetrahedron inscribed in S, and such that the center P is in the interior of the tetrahedron. Derive from each of these cases a contradiction with the assumptions of the theorem.

For more problems on Helly's theorem, and convexity in its many ramifications, see the excellent book **[4]** of Yaglom and Boltyanskii. See **[2**, §§6 and 7**]** for applications of Helly's theorem in R^n to Chebyshev's approximation problem. In [**4**, footnote on p. 16] we find that the results of Rademacher and Schoenberg in [**2**] were anticipated by some 12 years by the Russian mathematician L. G. Shnirelman.

References

[**1**] E. Helly, *Über Mengen konvexer Körper mit gemeinschaftlichen Punkten*, Jahresber. Deutsch. Math.-Verein., 32 (1923) 175–176.

[**2**] H. Rademacher and I. J. Schoenberg, *Convex domains and Chebyshev's approximation problem*, Canad. J. Math., 2 (1950) 245–256.

[**3**] A. Tarski, *A decision method for elementary algebra and geometry*, Rand, 1948.

[**4**] I. Yaglom and V. Boltyanskii, *Convex Figures*, Holt, Rinehart and Winston, New York, 1961.

CHAPTER 6

THE FINITE FOURIER SERIES I: THE GEOMETRIC AND ALGEBRAIC FOUNDATIONS

1. Introduction. Our MTEs are independent of each other and can be read in any order, an exception being the four MTEs devoted to the finite Fourier series (abbreviated to f.F.s.). Here we introduce the f.F.s. and develop some of its properties. The other three essays are independent of each other and deal with applications of an algebraic or geometric nature.

We assume the reader to be familiar with complex numbers, the operations with them, and their geometric interpretations. These matters are found in any introduction to complex analysis. Actually we need only the precalculus developments; for their far-reaching significance, see Schwerdtfeger's book [**3**]. We also use the rudiments of linear algebra. Eigenvalues of matrices, although they appear implicitly in §5, are not mentioned as such.

In the present introduction we deal with two easy problems on polygons in the complex plane. This subject was thoroughly studied by Jesse Douglas, of Plateau problem fame, in his article [**1**]; also by E. Kasner and some of his students (for references see [**1**]). We begin with Kasner's problem.

A. Let

$$\Pi = (z_0, z_1, \ldots, z_{k-1}) \tag{1.1}$$

be a closed k-gon in the plane, and let

$$\Pi' = (z'_0, z'_1, \ldots, z'_{k-1}) \tag{1.2}$$

be the k-gon having as vertices the midpoints of the successive sides of Π; hence

$$z'_0 = \tfrac{1}{2}(z_0 + z_1), \quad z'_1 = \tfrac{1}{2}(z_1 + z_2), \ldots, z'_{k-1} = \tfrac{1}{2}(z_{k-1} + z_0). \tag{1.3}$$

We now try to invert this relationship, a frequent source of fruitful problems, and propose

Problem 1. *If the midpoint polygon* Π' *is preassigned, to construct the original polygon* Π.

Algebraically, this is not much of a problem, for all we have to do is to try to invert the system (1.3) of k linear equations. Nevertheless the following geometric approach seems instructive.

We start from the point z and reflect it with respect to the point z_0'; this means that we pick $z^{(1)}$ such that z_0' is the midpoint of the segment from z to $z^{(1)}$, whence

$$z^{(1)} = 2z_0' - z.$$

We now reflect $z^{(1)}$ in z_1' to obtain

$$z^{(2)} = 2z_1' - z^{(1)}.$$

After k such reflections in the successive vertices of Π' we reach the point $z^{(k)}$ (Figure 6.1). To obtain the polygon Π we are to find z such that $z = z^{(k)}$.

What is the mapping $z \to z^{(k)}$?

Its nature depends on the parity of k.

Case 1: *k is odd*, $k=3$, say (Figure 6.2). Eliminating $z^{(1)}$ and $z^{(2)}$ from three relations

$$z^{(1)} = 2z_0' - z, \quad z^{(2)} = 2z_1' - z^{(1)}, \quad z^{(3)} = 2z_2' - z^{(2)}$$

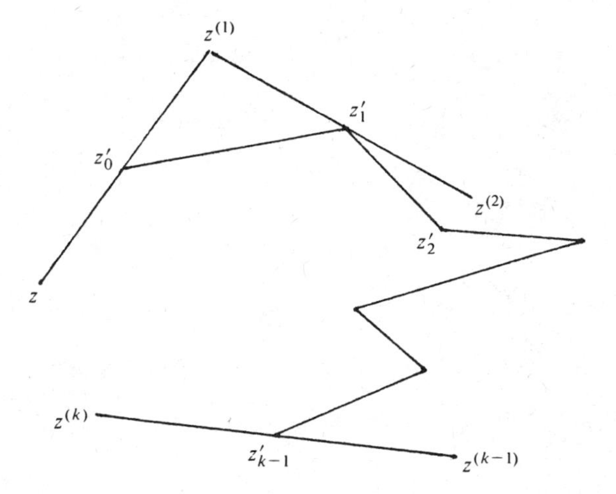

FIG. 6.1

we obtain

$$z^{(3)} = 2(z_2' - z_1' + z_0') - z, \tag{1.4}$$

showing that the mapping $z \to z^{(3)}$ is a reflection in a certain point z_0. However, we need not construct z_0 using (1.4), because it is already furnished by our construction of Figure 6.2 as

$$z_0 = \tfrac{1}{2}(z + z^{(3)}).$$

Once that z_0 is known, we obtain from it the polygon Π by reflections.

Case 2: *k is even*, $k = 4$, say (Figure 6.3). Starting from z we now have four reflections

$$z^{(1)} = 2z_0' - z, \quad z^{(2)} = 2z_1' - z^{(1)}, \quad z^{(3)} = 2z_2' - z^{(2)}, \quad z^{(4)} = 2z_3' - z^{(3)},$$

and, eliminating as above, we find that $z \to z^{(4)}$ amounts to

$$z^{(4)} = 2(z_3' - z_2' + z_1' - z_0') + z. \tag{1.5}$$

This is a translation of the plane: The open quadrilateral line $zz^{(1)}z^{(2)}z^{(3)}z^{(4)}$ will close if and only if

$$z_0' + z_2' = z_1' + z_3',$$

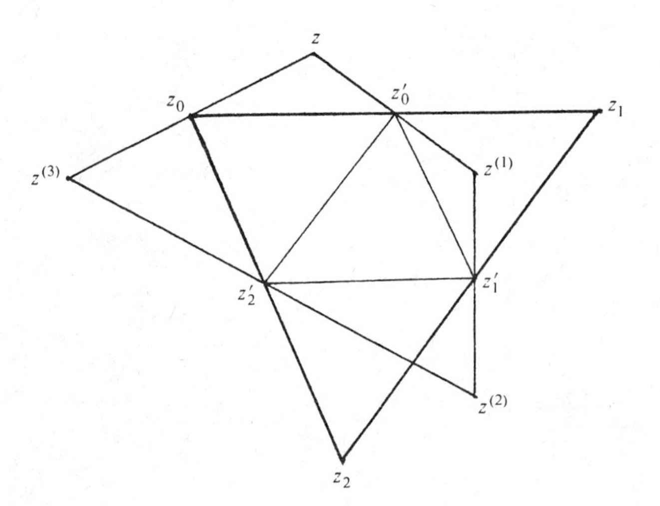

FIG. 6.2

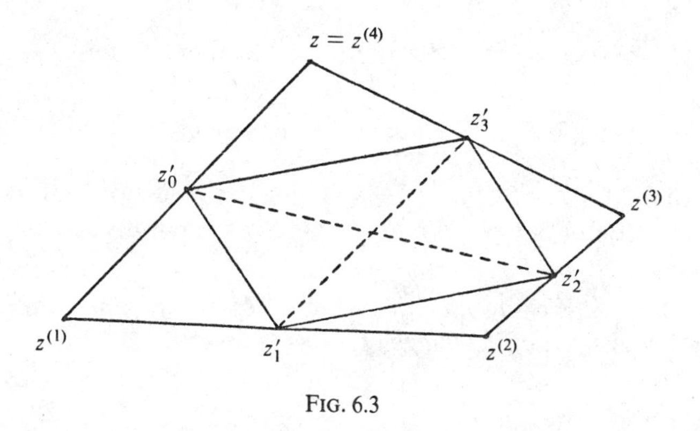

FIG. 6.3

and if this holds, then (1.5) shows that $z^{(4)} = z$ no matter what point z we start from.

We have established the following solution of Problem 1.

THEOREM OF KASNER. *If k is odd, then the polygon Π exists uniquely, and Figure 6.2 shows how to construct it.*

If k is even, $k = 2q$, then Π exists if and only if the given polygon Π' is such that the relation

$$z'_0 + z'_2 + \cdots + z'_{2q-2} = z'_1 + z'_3 + \cdots + z'_{2q-1} \tag{1.6}$$

holds, and then there are infinitely many solutions: the vertex z_0 may be picked at will.

Geometrically, the relation (1.6) means that the vertices of Π' having even subscripts, and those having odd subscripts, should have the same centroid.

B. The simplest and perhaps most important point-to-point transformation is the mapping $(x, y) \to (x', y')$, where

$$x' = ax + by$$

$$y' = cx + dy. \tag{1.7}$$

Except that it maps the origin into itself, (1.7) contains as special cases the rigid motions of the plane into itself, reflections, and shearing deformations. (1.7) is called an *affine mapping*, a name due to Euler. It maps straight lines into straight lines and preserves the parallelism of lines. If A, B, C are collinear points with images A', B', C', then $AB/BC = A'B'/B'C'$.

Let

$$\Pi = (z_0, z_1, \ldots, z_{k-1}) \tag{1.8}$$

be the regular k-gon having as vertices the kth roots of unity

$$z_\nu = e^{(2\pi i/k)\nu} \qquad (\nu=0,1,\ldots,k-1). \tag{1.9}$$

See Figure 6.4 for $k=5$, where Π is the regular pentagon with $\omega=e^{2\pi i/5}$. The following definition plays an important role in our discussion.

DEFINITION 1. *We say that the k-gon* (1.8) *is an affine regular k-gon, provided that* Π *is an affine image of a regular k-gon.*

We may now state our

Problem 2. *Let C and C′ be two concentric circles with the common center at O. In C we inscribe a regular* 5*-gon*

$$\Pi_u = (u_0, u_1, u_2, u_3, u_4),$$

and in C′ we inscribe a regular 5*-gon*

$$\Pi_v = (v_0, v_1, v_2, v_3, v_4),$$

such that its vertices v_ν *describe C′ clockwise, in the direction opposite to the direction of the* u_ν. *Finally, let*

$$z_\nu = u_\nu + v_\nu \qquad (\nu=0,1,2,3,4). \tag{1.10}$$

Show that the pentagon

$$\Pi = (z_0, z_1, z_2, z_3, z_4) \tag{1.11}$$

is affine regular.

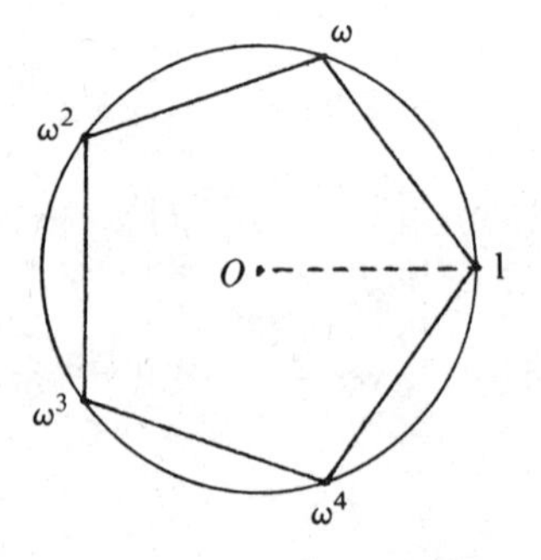

FIG. 6.4

Indeed, using the 5th root of unity

$$\omega = e^{2\pi i/5} = \cos\frac{2\pi}{5} + i\sin\frac{2\pi}{5}, \tag{1.12}$$

we have

$$u_\nu = u_0\omega^\nu,\ v_\nu = v_0\omega^{-\nu}\quad (\nu=0,1,\ldots,4),$$

and now (1.10) shows that

$$z_\nu = u_0\omega^\nu + v_0\omega^{-\nu}\quad (\nu=0,\ldots,4). \tag{1.13}$$

Writing

$$u_0 = A + iB,\ v_0 = C + iD,\quad (A, B, C, D \text{ real})$$

(1.13) shows that

$$z_\nu = (A+iB)\left(\cos\frac{2\pi\nu}{5} + i\sin\frac{2\pi\nu}{5}\right) + (C+iD)\left(\cos\frac{2\pi\nu}{5} - i\sin\frac{2\pi\nu}{5}\right).$$

Writing

$$z_\nu = x_\nu + iy_\nu \tag{1.14}$$

we obtain

$$x_\nu = (A+C)\cos\frac{2\pi\nu}{5} + (-B+D)\sin\frac{2\pi\nu}{5}$$

$$y_\nu = (B+D)\cos\frac{2\pi\nu}{5} + (A-C)\sin\frac{2\pi\nu}{5}.$$

Introducing the abbreviations

$$a = A + C,\quad b = -B + D,\quad c = B + D,\quad d = A - C,$$

we finally obtain

$$x_\nu = a\cos\frac{2\pi\nu}{5} + b\sin\frac{2\pi\nu}{5},$$

$$y_\nu = c\cos\frac{2\pi\nu}{5} + d\sin\frac{2\pi\nu}{5},\qquad (\nu=0,\ldots,4). \tag{1.15}$$

These equations show that the *pentagon* (1.11) *of Figure* 6.5 *is an affine*

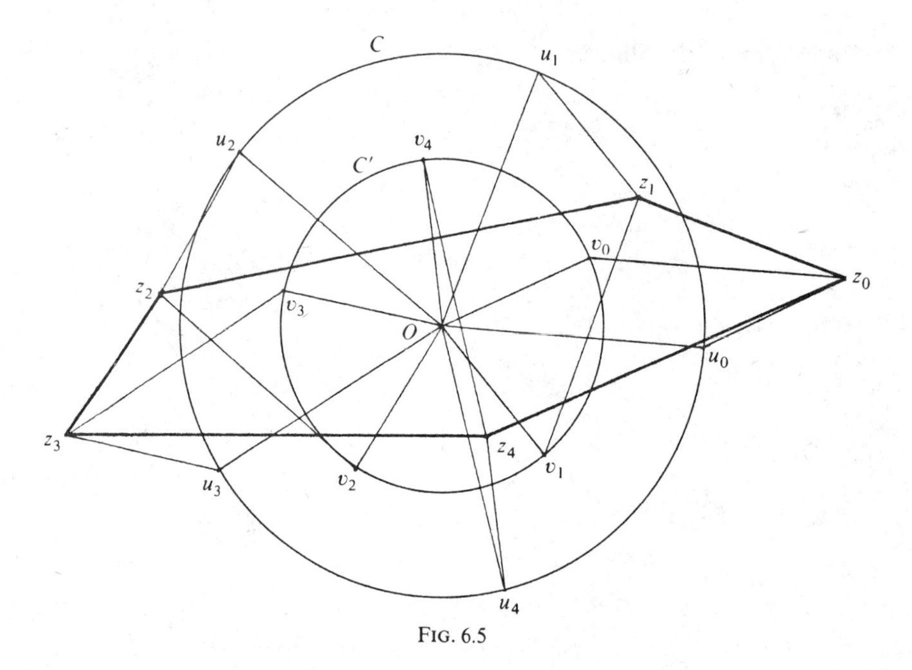

FIG. 6.5

image of the regular pentagon of Figure 6.4. By Definition 1, Π is affine regular.

Problem 2 and its proof generalize from 5 to any value of k. In fact we can even let $k \to \infty$ and conclude that the curve described by the point

$$z(t) = u_0 e^{it} + v_0 e^{-it} \qquad (0 \leq t \leq 2\pi)$$

is an affine image of the circumference, hence an *ellipse*.

An exercise. Use the formulae above to show that the affine transformation (1.15) is singular; i.e., the pentagon Π has its vertices on a straight line through O, if and only if

$$|u_0| = |v_0|,$$

which means that the circles C and C' coincide. (We recommend that the reader should make a good drawing for the case $C = C'$, similar to Figure 6.5.)

The last and most important topic to be dealt with in this introduction are some properties of the kth roots of unity defined by

$$\omega_\nu = e^{2\pi i\nu/k} \qquad (\nu = 0, 1, \ldots, k-1), \tag{1.16}$$

and especially their *orthogonality*. Let

$$\Omega_k = \begin{Vmatrix} 1 & 1 & \cdots & 1 \\ \omega_0 & \omega_1 & \cdots & \omega_{k-1} \\ \omega_0^2 & \omega_1^2 & \cdots & \omega_{k-1}^2 \\ \vdots & \vdots & & \vdots \\ \omega_0^{k-1} & \omega_1^{k-1} & \cdots & \omega_{k-1}^{k-1} \end{Vmatrix} = \| \omega^{\alpha\beta} \| \qquad (\alpha, \beta = 0, \dots, k-1), \tag{1.17}$$

where

$$\omega = e^{2\pi i/k}, \tag{1.18}$$

be the so-called Vandermonde matrix of the k numbers (1.16). The last expression (1.17) shows that Ω_k is a symmetric matrix, hence identical with its transpose Ω_k^T. We write explicitly a few special cases

$$\Omega_2 = \begin{Vmatrix} 1 & 1 \\ 1 & -1 \end{Vmatrix}, \quad \Omega_3 = \begin{Vmatrix} 1 & 1 & 1 \\ 1 & \omega & \omega^2 \\ 1 & \omega^2 & \omega \end{Vmatrix}, \quad \Omega_4 = \begin{Vmatrix} 1 & 1 & 1 & 1 \\ 1 & i & -1 & -i \\ 1 & -1 & 1 & -1 \\ 1 & -i & -1 & i \end{Vmatrix}. \tag{1.19}$$

The most compact form of the orthogonality property is the matrix relation

$$\Omega_k \cdot \overline{\Omega}_k^T = kI \qquad (I \text{ is the unit matrix}), \tag{1.20}$$

where $\overline{\Omega}_k$ is the matrix of the complex conjugates of the elements of Ω_k. Explicitly this means that

$$\omega_0^\alpha \overline{\omega}_0^\beta + \omega_1^\alpha \overline{\omega}_1^\beta + \cdots + \omega_{k-1}^\alpha \overline{\omega}_{k-1}^\beta = \begin{cases} k & \text{if} \quad \alpha = \beta, \\ 0 & \text{if} \quad \alpha \neq \beta. \end{cases} \tag{1.21}$$

This is clear if $\alpha = \beta$, since $\omega_\nu^\alpha \overline{\omega}_\nu^\beta = (\omega_\nu \overline{\omega}_\nu)^\alpha = |\omega_\nu|^{2\alpha} = 1$, and each term of (1.21) is $=1$.

Let us now assume that $\alpha \neq \beta$. Recalling the identity

$$1 + x + x^2 + \cdots + x^{k-1} = (x^k - 1)/(x-1) \qquad (x \neq 1), \tag{1.22}$$

the left-hand side of (1.21) becomes

$$\sum_{\gamma=0}^{k-1} \omega_\gamma^\alpha \overline{\omega}_\gamma^\beta = \sum_\gamma \omega^{\alpha\gamma} \omega^{-\beta\gamma} = \sum_{\gamma=0}^{k-1} \omega^{(\alpha-\beta)\gamma}.$$

Setting $x=\omega^{\alpha-\beta}$ and observing that $-(k-1)\leqslant\alpha-\beta\leqslant k-1$, we have $x\neq 1$, and now (1.22) shows that the last sum becomes

$$(\omega^{(\alpha-\beta)k}-1)/(\omega^{\alpha-\beta}-1)=0$$

because $\omega^{(\alpha-\beta)k}=(\omega^k)^{\alpha-\beta}=1$.

An exercise. Verify directly the relation (1.21), or (1.20), on the special matrices (1.19).

A last important remark: We know that matrix relations, like (1.20), give rise to the similar relations between the determinants of the factors. Since $\det\overline{\Omega}_k^T=\overline{\det\Omega_k}$, (1.20) shows that $|\det\Omega_k|^2=k^k$; hence

$$|\det\Omega_k| = k^{k/2}\neq 0. \tag{1.23}$$

This will be used in the next section.

2. The Finite Fourier Series. The k real or complex quantities

$$z_0, z_1, \ldots, z_{k-1} \tag{2.1}$$

being preassigned, we determine the k quantities

$$\zeta_0, \zeta_1, \ldots, \zeta_{k-1} \tag{2.2}$$

from the equations

$$z_\nu = \zeta_0+\zeta_1\omega_\nu+\zeta_2\omega_\nu^2+\cdots+\zeta_{k-1}\omega_\nu^{k-1} \qquad (\nu=0,\ldots,k-1). \tag{2.3}$$

This is a system of k simultaneous equations in the unknowns ζ_ν. The matrix of this system is the transpose Ω_k^T of the matrix (1.17). By (1.23) the system (2.3) is *non-singular* and therefore determines the ζ_ν uniquely. These could be determined from (2.3) by Cramer's rule. Fortunately, the orthogonality property allows us to solve the system (2.3), or

$$z_\nu = \sum_{\beta=0}^{k-1}\zeta_\beta\omega_\nu^\beta, \tag{2.4}$$

very easily: For if we multiply the equation (2.4) by $\overline{\omega}_\alpha^\nu=\overline{\omega}_\nu^\alpha$, and sum over all ν, we find on summing the double sum first with respect to ν that

$$\sum_{\nu=0}^{k-1} z_\nu\overline{\omega}_\alpha^\nu = \sum_{\beta,\nu}\zeta_\beta\omega_\nu^\beta\overline{\omega}_\nu^\alpha = \sum_{\beta=0}^{k-1}\zeta_\beta\sum_{\nu=0}^{k-1}\omega_\nu^\beta\overline{\omega}_\nu^\alpha = \zeta_\alpha\cdot k$$

by the orthogonality relation (1.21). The unknowns ζ_α of the system (2.3) are therefore explicitly given by

$$\zeta_\alpha = \frac{1}{k}\left(z_0 + z_1\bar{\omega}_\alpha + z_2\bar{\omega}_\alpha^2 + \cdots + z_{k-1}\bar{\omega}_\alpha^{k-1}\right). \tag{2.5}$$

The representation of the z_ν in the form (2.3) is called *the finite Fourier series* (abbreviated to f.F.s.) of the sequence (2.1). The coefficients ζ_α, given by (2.5), are called *the finite Fourier coefficients* of the z_ν.

EXAMPLES. 1. Let $k=2$ and

$$(z_0, z_1) = (1, -1). \tag{2.6}$$

From (2.5) we find that

$$\zeta_0 = \tfrac{1}{2}(z_0 + z_1) = 0, \quad \zeta_1 = \tfrac{1}{2}(z_0 - z_1) = 1,$$

and the f.F.s. (2.3) of our sequence becomes

$$z_\nu = (-1)^\nu. \tag{2.7}$$

Notice how the f.F.s. (2.7) extends the given sequence (2.6) to an infinite sequence of period 2: $1, -1, 1, -1, \ldots$. The same remark applies to the general sequence (2.1). *The* f.F.s. (2.3) *defines z_ν, for all integers ν, as an infinite periodic sequence of period k.*

2. Let $k=4$ and

$$(z_0, z_1, z_2, z_3) = (1, 1, -1, 1). \tag{2.8}$$

Here $\bar{\omega}_\alpha = (-i)^\alpha$ and we readily find that

$$(\zeta_0, \zeta_1, \zeta_2, \zeta_3) = \left(\tfrac{1}{2}, \tfrac{1}{2}, -\tfrac{1}{2}, \tfrac{1}{2}\right). \tag{2.9}$$

The f.F.s. of (2.8) is therefore

$$z_\nu = \tfrac{1}{2} + \tfrac{1}{2}i^\nu - \tfrac{1}{2}i^{2\nu} + \tfrac{1}{2}i^{3\nu}$$

or

$$z_\nu = \tfrac{1}{2} + \tfrac{1}{2}i^\nu - \tfrac{1}{2}(-1)^\nu + \tfrac{1}{2}i^{-\nu}. \tag{2.10}$$

Notice that the sum of the second and fourth terms is

$$\frac{1}{2}(i^\nu + i^{-\nu}) = \frac{1}{2}(e^{\nu\pi i/2} + e^{-\nu\pi i/2}) = \cos\left(\frac{\pi}{2}\nu\right),$$

and (2.10) becomes

$$z_\nu = \frac{1}{2}\left(1-(-1)^\nu\right)+\cos\frac{\pi\nu}{2}. \tag{2.11}$$

REMARKS. 1. Suppose you were asked to find the νth term of the infinite sequence $1,-1,1,-1,\ldots$ of Example 1. You would have certainly guessed the answer to be given by (2.7). It would not be equally easy to guess that the νth term of the sequence

$$1,1,-1,1,1,1,-1,1,\ldots$$

is given by (2.11). *The* f.F.s *answers the question automatically.*

2. Please skip this if you find it tedious, but I can't refrain from mentioning that the f.F.s. (2.11) for the sequence (2.8) can also be obtained by another method equivalent to the f.F.s. for our particular problem: *Finding x_ν for periodic sequences*. I mean the powerful method of *generating functions*, of which Euler, Laplace, and J. J. Sylvester were masters. All we have to do is to *sum* the power series

$$f(x) = 1+x-x^2+x^3+x^4+x^5-x^6+x^7+\cdots \qquad (|x|<1), \tag{2.12}$$

and afterwards *expanding $f(x)$ in powers of x*. You may argue that we already have in (2.12) the expansion of $f(x)$ in powers of x, but wait and see. Evidently

$$f(x) = \sum_0^\infty x^\nu - 2x^2\sum_0^\infty x^{4\nu} = \frac{1}{1-x}-\frac{2x^2}{1-x^4} = \frac{1}{1-x}+\frac{1}{x^2-1}+\frac{1}{x^2+1}$$

$$= \frac{1}{1-x}-\frac{1}{(1-x)(1+x)}+\frac{1}{(1-ix)(1+ix)}$$

$$= \frac{1}{1-x}-\frac{1}{2}\frac{1}{1-x}-\frac{1}{2}\frac{1}{1+x}+\frac{1}{2}\frac{1}{1-ix}+\frac{1}{2}\frac{1}{1+ix}$$

$$= \frac{1}{2}\sum_0^\infty x^\nu - \frac{1}{2}\sum_0^\infty(-1)^\nu x^\nu + \frac{1}{2}\sum_0^\infty i^\nu x^\nu + \frac{1}{2}\sum_0^\infty(-i)^\nu x^\nu,$$

$$f(x) = \sum_0^\infty \frac{1}{2}\left(1-(-1)^\nu+i^\nu+i^{-\nu}\right)x^\nu$$

which agrees with (2.10). We have used the complex poles $\pm i$ of $f(x)$.

3. A Geometric Interpretation of the Finite Fourier Series. Using column vectors we may write the relations (2.3) in the form

$$\begin{Vmatrix} z_0 \\ z_1 \\ \vdots \\ z_{k-1} \end{Vmatrix} = \begin{Vmatrix} 1 & 1 & 1 & & 1 \\ 1 & \omega_1 & \omega_1^2 & \cdots & \omega_1^{k-1} \\ \vdots & & & & \\ 1 & \omega_{k-1} & \omega_{k-1}^2 & \cdots & \omega_{k-1}^{k-1} \end{Vmatrix} \cdot \begin{Vmatrix} \zeta_0 \\ \zeta_1 \\ \vdots \\ \zeta_{k-1} \end{Vmatrix}. \tag{3.1}$$

This shows that the left-hand vector is a linear combination of the columns of Ω_k, the coefficients being the ζ_ν. It is instructive to consider the set of k polygons defined by the columns of Ω_k:

$$\Pi_0 = (1,1,1,\ldots,1),$$

$$\Pi_1 = (1,\omega_1,\omega_2,\ldots,\omega_{k-1}),$$

$$\Pi_2 = (1,\omega_1^2,\omega_2^2,\ldots,\omega_{k-1}^2),$$

$$\vdots$$

$$\Pi_{k-1} = (1,\omega_1^{k-1},\omega_2^{k-1},\ldots,\omega_{k-1}^{k-1}). \tag{3.2}$$

Observe the shapes of these polygons: Π_0 is the point 1 repeated k times. Π_1 is our basic regular k-gon. However, Π_2 is obtained by starting from 1 and taking every other vertex of Π_1 until we have a total of k points. For $k=5$, Π_2 is the star-shaped pentagon of Figure 6.6. The vertices of Π_ν are

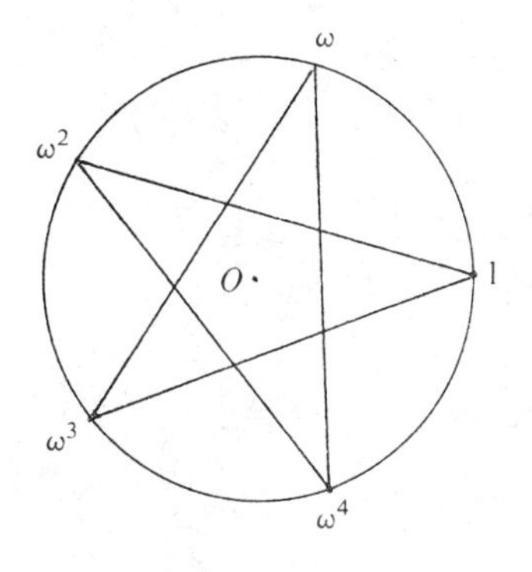

FIG. 6.6

obtained by taking every νth vertex of Π_1. In particular, from $\omega_\nu^{k-1} = \omega_\nu^{-1}$, we see that Π_{k-1} is again the polygon Π_1 taken in the clockwise direction.

The relation (3.1) shows that we may represent the polygon

$$\Pi = (z_0, z_1, \ldots, z_{k-1}) \tag{3.3}$$

in terms of the polygons (3.2) as

$$\Pi = \zeta_0 \Pi_0 + \zeta_1 \Pi_1 + \zeta_2 \Pi_2 + \cdots + \zeta_{k-1} \Pi_{k-1}. \tag{3.4}$$

We may say that *the* f.F.s. (2.3) *analyzes* Π *in terms of the k basic polygons* (3.2).

The expansion (2.3), or

$$z_\nu = \zeta_0 + \zeta_1 \omega_\nu + \zeta_2 \omega_\nu^2 + \cdots + \zeta_{k-2} \omega_\nu^{k-2} + \zeta_{k-1} \omega_\nu^{k-1}, \tag{3.5}$$

is sometimes called the *complex* f.F.s. To explain the reason for this name we begin by folding (3.5) to half its length by grouping its terms as follows:

$$z_\nu = \zeta_0 + \left(\zeta_1 \omega_\nu + \zeta_{k-1} \omega_\nu^{k-1}\right) + \left(\zeta_2 \omega_\nu^2 + \zeta_{k-2} \omega_\nu^{k-2}\right) + \cdots. \tag{3.6}$$

How this expansion terminates depends on the parity of k.

If k is *odd*, $k = 2n+1$, we obtain

$$z_\nu = \zeta_0 + \left(\zeta_1 \omega_\nu + \zeta_{-1} \omega_\nu^{-1}\right) + \left(\zeta_2 \omega_\nu^2 + \zeta_{-2} \omega_\nu^{-2}\right) + \cdots + \left(\zeta_n \omega_\nu^n + \zeta_{-n} \omega_\nu^{-n}\right). \tag{3.7}$$

If k is *even*, $k = 2n$, then (3.6) becomes

$$z_\nu = \zeta_0 + \left(\zeta_1 \omega_\nu + \zeta_{-1} \omega_\nu^{-1}\right) + \cdots + \left(\zeta_{n-1} \omega_\nu^{n-1} + \zeta_{-(n-1)} \omega_\nu^{-(n-1)}\right) + \zeta_n \omega_\nu^n. \tag{3.8}$$

It is sometimes convenient to write (3.8) as

$$z_\nu = \zeta_0 + \cdots + \left(\zeta_{n-1} \omega_\nu^{n-1} + \zeta_{-(n-1)} \omega_\nu^{-(n-1)}\right) + \tfrac{1}{2} \zeta_n \omega_\nu^n + \tfrac{1}{2} \zeta_n \omega_\nu^{-n}, \tag{3.8$'$}$$

which is correct because $\omega_\nu^n = \omega_\nu^{-n}$ if $k = 2n$.

From the solution of Problem 2 of our Introduction we know that the term $(\zeta_1 \omega_\nu + \zeta_{-1} \omega_\nu^{-1})$, by itself, represents an *affine regular k-gon* A_1 having the origin as center (see Figure 6.7 and the relation (1.13) where $\zeta_1 = u_0$ and $\zeta_{-1} = v_0$). Similarly, the other terms of (3.7), or (3.8), represent various

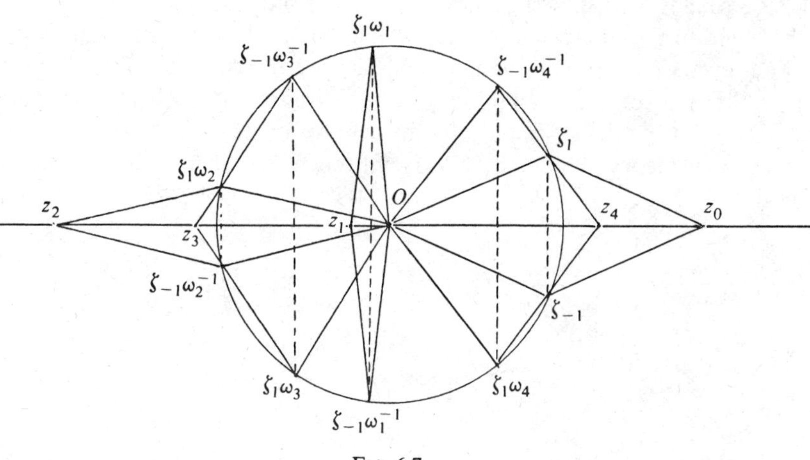

FIG. 6.7

affine regular (star-shaped) k-gons. Therefore our relations (3.7) and (3.8) may be interpreted as follows:

They show that the k-gon (3.3) *can be represented as a sum*

$$\Pi = \zeta_0\Pi_0 + A_1 + A_2 + \cdots + A_n \qquad \left(n = \left[\frac{k}{2}\right]\right) \tag{3.9}$$

of affine regular k-gons. It seems remarkable that this representation is *unique*, a fact that follows from the unicity of the f.F.s. (3.5).

At this point we need

THEOREM 1. *The numbers*

$$z_\nu \quad (\nu = 0,\ldots,k-1) \quad \textit{are real} \tag{3.10}$$

if and only if the f.F. *coefficients* ζ_ν *are imaginary conjugate in pairs, which means that*

$$\zeta_0 \quad \textit{is real, and} \quad \zeta_{-\nu} = \bar{\zeta}_\nu \qquad \left(\nu = 1,\ldots,\left[\frac{k}{2}\right]\right). \tag{3.11}$$

In Problem 1 at the end of this MTE the reader is asked to supply a proof of Theorem 1.

If the conditions for reality (3.11) hold, then also all individual terms of (3.7), or (3.8), are real for all ν. This is the reason that these expansions are called the *real finite Fourier series*. Figure 6.7 shows the reality of the vertices of the pentagon

$$z_\nu = \zeta_1\omega_\nu + \zeta_{-1}\omega_\nu^{-1} \qquad (\nu = 0,1,2,3,4)$$

if $k = 5$ and $\zeta_{-1} = \bar{\zeta}$.

4. The Analogue of the Parseval Relation. For purposes of motivation we use here and in the next section terms which are familiar from the theory of the Fourier series, a fact which in no way presupposes a knowledge of Fourier series theory. The Fourier series requires limit processes and was fundamental in the development of Analysis; for instance, it led G. Cantor to develop set theory, while the f.F.s. is entirely elementary.

Besides the f.F.s.

$$z_\nu = \sum_{\alpha=0}^{k-1} \zeta_\alpha \omega_\nu^\alpha \qquad (\nu=0,\dots,k-1), \tag{4.1}$$

let

$$x_\nu = \sum_{\beta=0}^{k-1} \xi_\beta \omega_\nu^\beta \tag{4.2}$$

be the f.F.s. of a second sequence (x_ν). We form the "inner-product" of the two sequences and find

$$\sum_{\nu=0}^{k-1} z_\nu \bar{x}_\nu = \sum_{\alpha,\beta,\nu} \zeta_\alpha \omega_\nu^\alpha \bar{\xi}_\beta \bar{\omega}_\nu^\beta = \sum_{\alpha,\beta} \zeta_\alpha \bar{\xi}_\beta \sum_{\nu=0}^{k-1} \omega_\nu^\alpha \bar{\omega}_\nu^\beta,$$

and therefore, by the orthogonality relations (1.21),

$$\sum_{\nu=0}^{k-1} z_\nu \bar{x}_\nu = k \sum_{\alpha=0}^{k-1} \zeta_\alpha \bar{\xi}_\alpha.$$

In particular, if $x_\nu = z_\nu$ for all ν, we find that

$$\sum_{\nu=0}^{k-1} |z_\nu|^2 = k \cdot \sum_{\alpha=0}^{k-1} |\zeta_\alpha|^2. \tag{4.3}$$

This is the *finite Parseval relation* which we shall use repeatedly.

Exercise. Verify directly that (4.3) holds for the sequence (2.8) and its f.F. coefficients (2.9).

5. The Convolution of Sequences. Let again (z_ν) and (x_ν) be the two sequences (4.1) and (4.2). Out of them we form a new sequence (z'_ν) defined by the equations

$$z'_\nu = \sum_{\alpha=0}^{k-1} z_\alpha x_{\nu-\alpha} \qquad (\nu=0,\dots,k-1), \tag{5.1}$$

where we regard both sequences as being periodic with period k, i.e., $x_{-1}=x_{k-1}$, $x_{-2}=x_{k-2}$, and so forth. The operation (5.1) is called *the convolution of the sequences* (z_ν) *and* (x_ν), and denoted symbolically by

$$(z'_\nu) = (z_\nu)*(x_\nu).$$

Notice that it is commutative; hence $(z_\nu)*(x_\nu)=(x_\nu)*(z_\nu)$. Explicitly the relations (5.1) are

$$\begin{aligned}
z'_0 &= z_0x_0 + z_1x_{k-1} + z_2x_{k-2} + \cdots + z_{k-1}x_1 \\
z'_1 &= z_0x_1 + z_1x_0 + z_2x_{k-1} + \cdots + z_{k-1}x_2 \\
&\vdots \\
z'_{k-1} &= z_0x_{k-1} + z_1x_{k-2} + z_2x_{k-3} + \cdots + z_{k-1}x_0. \qquad (5.2)
\end{aligned}$$

What are the f.F. *coefficients* (ζ'_ν) *of the convolution* (5.1)? As before we denote by (ξ_ν) the coefficients of (x_ν) and find

$$\begin{aligned}
z'_\nu &= \sum_\alpha z_\alpha x_{\nu-\alpha} = \sum_{\lambda,\mu,\alpha} \zeta_\lambda\omega_\alpha^\lambda\xi_\mu\omega_{\nu-\alpha}^\mu = \sum_{\lambda,\mu,\alpha} \zeta_\lambda\xi_\mu\omega_\nu^\mu\omega_\alpha^\lambda\overline{\omega}_\alpha^\mu \\
&= \sum_{\lambda,\mu} \zeta_\lambda\xi_\mu\omega_\nu^\mu \sum_{\alpha=0}^{k-1} \omega_\alpha^\lambda\overline{\omega}_\alpha^\mu
\end{aligned}$$

and the orthogonality relations (1.21) show that

$$z'_\nu = \sum_{\lambda=0}^{k-1} (k\zeta_\lambda\xi_\lambda)\omega_\nu^\lambda \qquad (\nu=0,\ldots,k-1).$$

This being the f.F.s. of (z'_ν), we conclude that *the coefficients of the convolution* $(z'_\nu)=(z_\nu)*(x_\nu)$ *are given by*

$$\zeta'_\nu = \zeta_\nu\xi_\nu k \qquad (\nu=0,\ldots,k-1). \qquad (5.3)$$

We want to express this remarkably simple result in a more convenient form. We may look at the convolution operation (5.1) as the linear transformation (5.2), from the variables (z_ν) to the new variables (z'_ν). The transformation (5.2) may be described as a *cyclic transformation*, because the successive rows of its matrix are obtained from its first row by successive *cyclic permutations*. The equations (5.3) show that the f.F. coefficients of

(z'_ν) are obtained by multiplying the coefficients ζ_ν by the factor

$$\xi_\nu k = x_0 + x_1\bar{\omega}_\nu + x_2\bar{\omega}_\nu^2 + \cdots + x_{k-1}\bar{\omega}_\nu^{k-1},$$

which we prefer to write in the equivalent form

$$\xi_\nu k = x_0 + x_{k-1}\omega_\nu + x_{k-2}\omega_\nu^2 + \cdots + x_1\omega_\nu^{k-1}, \tag{5.4}$$

since $\bar{\omega}_\nu^\alpha = \omega_\nu^{k-\alpha}$. If we change our notations by writing

$$a_0 = x_0, \quad a_1 = x_{k-1}, \quad a_2 = x_{k-2}, \dots, a_{k-1} = x_1$$

then (5.4) becomes

$$\xi_\nu k = a_0 + a_1\omega_\nu + a_2\omega_\nu^2 + \cdots + a_{k-1}\omega_\nu^{k-1}. \tag{5.5}$$

This and (5.3) establish the following

THEOREM 2. *If (ζ_ν) are the finite Fourier coefficients of the sequence (z_ν), and if we subject the (z_ν) to the cyclic transformation*

$$\begin{aligned} z'_0 &= a_0 z_0 + a_1 z_1 + a_2 z_2 + \cdots + a_{k-1} z_{k-1} \\ z'_1 &= a_{k-1} z_0 + a_0 z_1 + a_1 z_2 + \cdots + a_{k-2} z_{k-1} \\ &\ \vdots \\ z'_{k-1} &= a_1 z_0 + a_2 z_1 + a_3 z_2 + \cdots + a_0 z_{k-1}, \end{aligned} \tag{5.6}$$

then the finite Fourier coefficients of the new sequence (z'_ν) are

$$\zeta'_\nu = \zeta_\nu f(\omega_\nu), \tag{5.7}$$

where

$$f(z) = a_0 + a_1 z + \cdots + a_{k-1} z^{k-1}, \tag{5.8}$$

and will be referred to as the representative polynomial of the cyclic transformation (5.6).

A last remark is the following. By the Parseval relation (4.3) we have

$$\sum |z'_\nu|^2 = k \cdot \sum |\zeta'_\nu|^2.$$

Substituting on the right-hand side the values (5.7), we obtain the

COROLLARY 1. *The relations* (5.6) *imply the identity*

$$\sum_{\nu=0}^{k-1} |z'_\nu|^2 = k \sum_{\nu=0}^{k-1} |f(\omega_\nu)|^2 |\zeta_\nu|^2, \tag{5.9}$$

where (ζ_ν) *are the finite Fourier coefficients of the* (z_ν), *and* $f(z)$ *is defined by* (5.8).

The relation (5.9) should be regarded as an identity in the variables (z_ν), the (ζ_ν) being given in terms of the (z_ν) by (2.5).

Problems

1. The f.F.s. is summarized by the two linear transformations (2.3) and (2.5) which are inverse to each other. Use them to establish Theorem 1 of §3.

2. (i) Let $\omega = \exp(2\pi i/3)$. Show that the triangle $T = (z_0, z_1, z_2)$ is a regular triangle if and only if one of the two relations

$$z_0 + z_1\omega + z_2\omega^2 = 0, \qquad z_0 + z_1\omega^2 + z_2\omega = 0,$$

holds.

(ii) Conclude that T is regular if and only if

$$z_0^2 + z_1^2 + z_2^2 - z_0z_1 - z_0z_2 - z_1z_2 = 0.$$

(Notice that $1 + \omega + \omega^2 = 0$.)

References

[1] Jesse Douglas, *Geometry of polygons in the complex plane*, J. Math. Phys., 19 (1940) 93–130.

[2] I. J. Schoenberg, *The finite Fourier series and elementary geometry*, Amer. Math. Monthly, 57 (1950) 390–404.

[3] Hans Schwerdtfeger, *Geometry of Complex Numbers*, University of Toronto Press, Toronto, 1962.

CHAPTER 7

THE FINITE FOURIER SERIES II: THE CASE OF THE TRIANGLE ($k=3$)

1. Introduction. We specialize the finite Fourier series of the previous chapter by assuming that $k=3$, so that we are given a triangle

$$T=(z_0,z_1,z_2).$$

As we do not use any of its more elaborate properties, we make this brief chapter independently readable by defining again the finite Fourier series for this special case.

We develop two applications: 1. The formulae of *Cardan* for the solution of the general cubic equation. 2. The theorem of *van den Berg* [**1**] giving the location of the zeros of the derivative of a cubic polynomial having as zeros the vertices of the triangle T. On the geometry of polynomials see Marden's book [**3**]. For a different discussion of the van den Berg theorem see Walsh's paper [**6**]. Further references will be given as needed.

2. The Finite Fourier Series of the Triangle. Let

$$T=(z_0,z_1,z_2) \tag{2.1}$$

be a triangle in the complex plane having the vertices z_0, z_1, z_2. Let $\omega=\exp(2\pi i/3)$ be the third root of unity, so that 1, ω, ω^2 are the roots of the equation $z^3-1=0$. A consequence is such relations as $\omega^4=\omega$, $\omega^2=\bar{\omega}$, $1+\omega+\omega^2=0$, a.s.f.

We look for the quadratic polynomial $P(z)=\zeta_0+\zeta_1 z+\zeta_2 z^2$ such that $P(1)=z_0$, $P(\omega)=z_1$, $P(\omega^2)=z_2$, hence satisfying

$$\begin{aligned} z_0 &= \zeta_0+\zeta_1+\zeta_2 \\ z_1 &= \zeta_0+\zeta_1\omega+\zeta_2\omega^2 \\ z_2 &= \zeta_0+\zeta_1\omega^2+\zeta_2\omega. \end{aligned} \tag{2.2}$$

This is an interpolation problem that is readily solved, provided that we first verify the matrix equation

$$\begin{Vmatrix} 1 & 1 & 1 \\ 1 & \omega & \omega^2 \\ 1 & \omega^2 & \omega \end{Vmatrix} \begin{Vmatrix} 1 & 1 & 1 \\ 1 & \omega^2 & \omega \\ 1 & \omega & \omega^2 \end{Vmatrix} = \begin{Vmatrix} 3 & 0 & 0 \\ 0 & 3 & 0 \\ 0 & 0 & 3 \end{Vmatrix}. \tag{2.3}$$

A consequence is that we obtain the ζ's of (2.2) in the form

$$\begin{aligned} 3\zeta_0 &= z_0 + z_1 + z_2 \\ 3\zeta_1 &= z_0 + z_1\omega^2 + z_2\omega \\ 3\zeta_2 &= z_0 + z_1\omega + z_2\omega^2. \end{aligned} \tag{2.4}$$

The equations (2.2) represent the *finite Fourier series of the sequence* z_0, z_1, z_2, while (2.4) are the expressions for its *finite Fourier coefficients* ζ_0, ζ_1, ζ_2.

The first equation (2.4) shows that $\zeta_0 = (z_0 + z_1 + z_2)/3$ is the centroid of the triangle T. Our formulae simplify if we place this centroid in the origin O of the complex plane. Assuming that

$$3\zeta_0 = z_0 + z_1 + z_2 = 0, \tag{2.5}$$

the equations (2.2) become

$$\begin{aligned} z_0 &= \zeta_1 + \zeta_2 \\ z_1 &= \zeta_1\omega + \zeta_2\omega^2 \\ z_2 &= \zeta_1\omega^2 + \zeta_2\omega. \end{aligned} \tag{2.6}$$

3. Cardan's Solution of the General Cubic Equation. We begin by deriving a few further relations between the z_ν and ζ_ν. From (2.4) we obtain that

$$9\zeta_1\zeta_2 = z_0^2 + z_1^2 + z_2^2 - z_0z_1 - z_0z_2 - z_1z_2. \tag{3.1}$$

Squaring (2.5) we have

$$z_0^2 + z_1^2 + z_2^2 + 2z_0z_1 + 2z_0z_2 + 2z_1z_2 = 0,$$

and subtracting this from (3.1) we see that

$$3\zeta_1\zeta_2 = -(z_0z_1 + z_0z_2 + z_1z_2). \tag{3.2}$$

Let us now assume that z_0, z_1, z_2 are the roots of the cubic equation

$$z^3 + pz + q = 0 \tag{3.3}$$

where the term in z^2 is missing in view of (2.5). Since $p = z_0 z_1 + z_0 z_2 + z_1 z_2$, we may write (3.2) as

$$3\zeta_1\zeta_2 = -p. \tag{3.4}$$

A further relation we derive from the identity $(z-1)(z-\omega)(z-\omega^2) = z^3 - 1$. Replacing z by $-z$ we obtain $(z+1)(z+\omega)(z+\omega^2) = z^3+1$, and setting here $z = \zeta_1/\zeta_2$ this becomes $(\zeta_1+\zeta_2)(\zeta_1+\omega\zeta_2)(\zeta_1+\omega^2\zeta_2) = \zeta_1^3+\zeta_2^3$. Multiplying the second factor on the left by ω, and the third by ω^2, we reach the equation

$$(\zeta_1+\zeta_2)(\zeta_1\omega+\zeta_2\omega^2)(\zeta_1\omega^2+\zeta_2\omega) = \zeta_1^3+\zeta_2^3.$$

Now (2.6) shows that this may be written as

$$z_0 z_1 z_2 = \zeta_1^3 + \zeta_2^3. \tag{3.5}$$

Since $z_0 z_1 z_2 = -q$, we finally have the two equations

$$\begin{aligned} \zeta_1^3 + \zeta_2^3 &= -q, \\ \zeta_1^3\zeta_2^3 &= -p^3/27, \end{aligned} \tag{3.6}$$

the second being obtained by cubing (3.4).

These equations show that ζ_1^3 and ζ_2^3 are the two roots of the so-called *quadratic resolvent*

$$x^2 + qx - \frac{p^3}{27} = 0, \tag{3.7}$$

and therefore

$$\zeta_1^3 = -\frac{q}{2} + \sqrt{\Delta}, \quad \zeta_2^3 = -\frac{q}{2} - \sqrt{\Delta}, \tag{3.8}$$

where

$$\Delta = \frac{q^2}{4} + \frac{p^3}{27}. \tag{3.9}$$

We therefore obtain the roots of the cubic (3.3) as follows: *Determine* ζ_1 *from the first equation* (3.8), *and then* ζ_2 *from* (3.4), *or*

$$\zeta_1\zeta_2 = -p/3. \tag{3.10}$$

As there are *two* ways of selecting $\sqrt{\Delta}$, and *three* ways of choosing ζ_1 as a root of the first equation (3.8), we get the roots z_0, z_1, z_2, by (2.6) in *six* different ways. These only result in the six permutations of the roots.

We assume the cubic equation (3.3) to be real. Clearly, the nature of its roots will depend on the sign of the discriminant Δ. There are three cases.

1. $\Delta>0$. The right-hand side of (3.8) is real and we choose the unique real cubic root ζ_1, when also (3.10) gives a real ζ_2. Our Figure 7.1 illustrates a case when $p>0$. From (2.6) we see that (3.3) *has one real root* z_0, *while* z_1 *and* z_2 *are imaginary conjugate.*

2. $\Delta<0$. It is clear that $p<0$. From (3.8) we gather that $|\zeta_1|=|\zeta_2|$. We pick any one of the roots ζ_1 of (3.8), and (3.10) shows that ζ_1 *and* ζ_2 *are imaginary conjugate*. From (2.6) we see that *all three roots* z_0, z_1, z_2 *are real.* This case is shown in Figure 7.2.

3. If $\Delta=0$, again we have $p<0$. Choosing for ζ_1 the real root of (3.8), we satisfy (3.10) with $\zeta_2=\zeta_1$. The result is that all roots are real with $z_1=z_2$.

Our solution of the cubic equation (3.3) is due to Nicolo Tartaglia (1535), even though it is usually attributed to H. Cardano. It represents one of the great achievements of the Italian Renaissance (see [**5**, 91–93]).

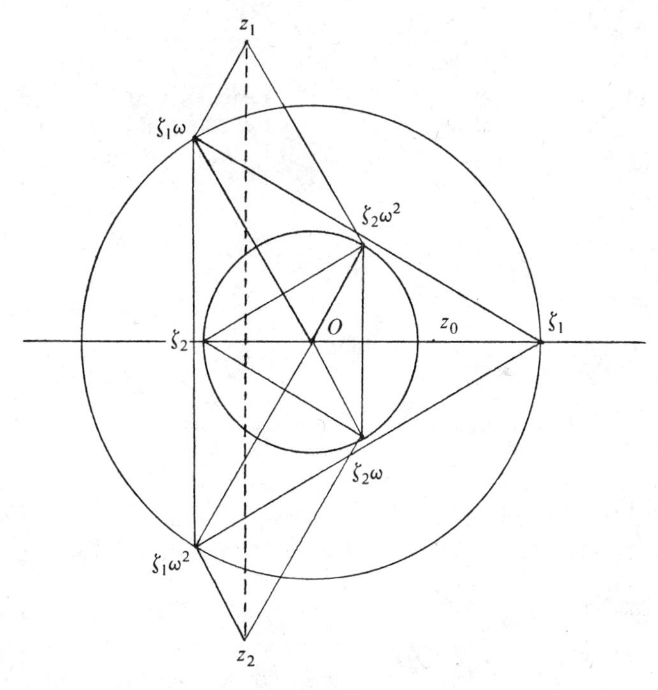

FIG. 7.1

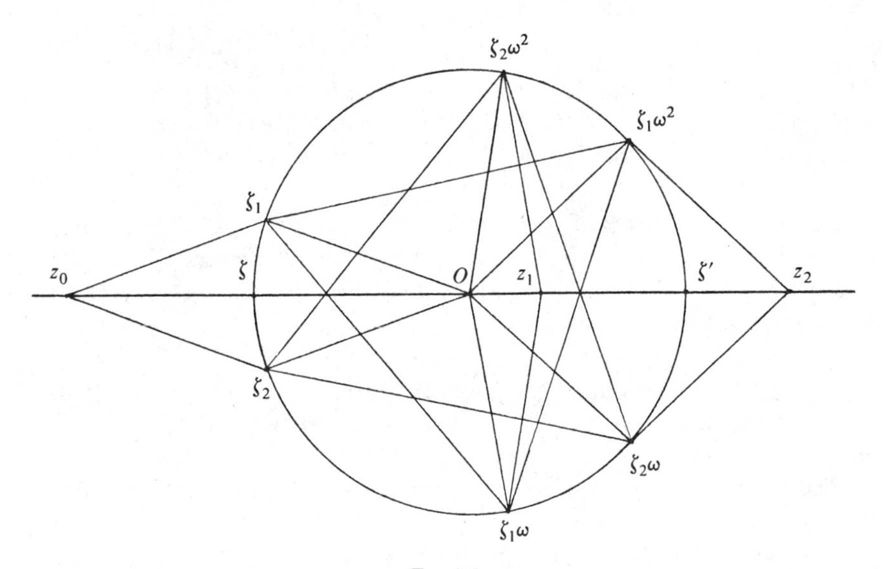

FIG. 7.2

4. On the Zeros of the Derivative of a Cubic Polynomial. Let

$$P(z) = (z - z_0)(z - z_1) \cdots (z - z_{k-1}) \tag{4.1}$$

be a polynomial of degree k having the zeros $z_0, \ldots, z_{k-1}$, which we plot in the complex plane. What can we say about the zeros $z_0', \ldots, z_{k-2}'$ of $P'(z)$? Forming the derivative $P'(z)$ and dividing it by $P(z)$ we get the remarkably simple identity

$$\frac{P'(z)}{P(z)} = \frac{1}{z - z_0} + \frac{1}{z - z_1} + \cdots + \frac{1}{z - z_{k-1}}. \tag{4.2}$$

If $P'(z_0') = 0$, and $z_0' \neq z_\nu$ for all ν, then

$$\frac{1}{z_0' - z_0} + \frac{1}{z_0' - z_1} + \cdots + \frac{1}{z_0' - z_{k-1}} = 0. \tag{4.3}$$

Using the usual vector representation of complex numbers we can interpret the equation (4.3) mechanically as follows: *If each of the points z_ν attracts the point z_0' with a force that is inverse proportional to the distance $|z_0' - z_\nu|$, then z_0', satisfying* (4.3), *is in a position of equilibrium*. This makes it intuitively plausible that z_0' belongs to the least convex polygon Π that contains the points $z_0, \ldots, z_n$. If $\Pi^{(\nu)}$ denotes the least convex polygon containing the zeros of $P^{(\nu)}(z)$, then we have the set-inclusions

$$\Pi \supset \Pi' \supset \Pi'' \supset \cdots \supset \Pi^{(k-1)}. \tag{4.4}$$

This is the theorem of Gauss-Lucas. See [**3**, Chapter 2] for a complete discussion of this subject.

It is remarkable that for the case $k=3$ of the cubic $P_3(z)$, when Π becomes the triangle

$$T = (z_0, z_1, z_2), \tag{4.5}$$

the location of the zeros of $P_3'(z)$ can be beautifully described. Let $\zeta_0 = 0$, ζ_1, ζ_2 be the finite Fourier coefficients of the vertices of the triangle (4.5), and let us write its finite Fourier series (2.6) in the form

$$z_\nu = \zeta_1 \omega^\nu + \zeta_2 \omega^{-\nu} \qquad (\nu = 0,1,2). \tag{4.6}$$

Alongside, we consider the curve

$$\mathfrak{S}: z = \zeta_1 e^{i\theta} + \zeta_2 e^{-i\theta} \qquad (0 \leqslant \theta \leqslant 2\pi). \tag{4.7}$$

The three points (4.6) are seen to lie on $\mathfrak{S}$ corresponding to the values of $\theta = 0$, $2\pi/3$, $4\pi/3$. Let us express $\mathfrak{S}$ also in real parametric form: Setting $z = x + iy$, $\zeta_1 = a_1 + ib_1$, $\zeta_2 = a_2 + ib_2$, and of course $e^{\pm i\theta} = \cos\theta \pm i\sin\theta$, (4.7) is found to be equivalent to the representation

$$\mathfrak{S}: \begin{array}{l} x = (a_1 + a_2)\cos\theta + (b_2 - b_1)\sin\theta, \\ y = (b_1 + b_2)\cos\theta + (a_1 - a_2)\sin\theta. \end{array} \tag{4.8}$$

It therefore appears that $\mathfrak{S}$ is the image of the unit circle

$$C: u = \cos\theta, \quad v = \sin\theta \tag{4.9}$$

by the affine transformation

$$A: \begin{array}{l} x = (a_1 + a_2)u + (b_2 - b_1)v \\ y = (b_1 + b_2)u + (a_1 - a_2)v. \end{array} \tag{4.10}$$

Also the triangle (4.5) is the image, by A, of the regular triangle

$$t = (1, \omega, \omega^2) \tag{4.11}$$

inscribed in C. At the same time we consider the circle *inscribed* in t. It has the radius $1/2$ and the representation

$$\tfrac{1}{2}C: u = \tfrac{1}{2}\cos\theta, \quad v = \tfrac{1}{2}\sin\theta. \tag{4.12}$$

Our Figure 7.3 has two parts, (a) and (b). In (b) we see the circles C and $\frac{1}{2}C$ and the triangle t. Figure 7.3(a) is the image of (b) by the affine transformation (4.10). The curve $\mathcal{E}$ is the image of C and is therefore an *ellipse*. Let E be the ellipse which is the image of $\frac{1}{2}C$. E is called the *Steiner Ellipse of the triangle* $T=(z_0, z_1, z_2)$, which is the image of t. Because in (b) the circle $\frac{1}{2}C$ is tangent to the sides of t and tangent to them at their midpoints, it follows that *the Steiner ellipse E is inscribed in T, touching its sides also at their midpoints*. Using obvious notations we also have

$$E = \tfrac{1}{2}\mathcal{E}. \tag{4.13}$$

We assume, of course, that the affine map (4.10) is not singular. As its determinant is

$$D = a_1^2 - a_2^2 + b_1^2 - b_2^2 = |\zeta_1|^2 - |\zeta_2|^2,$$

we assume that

$$|\zeta_1| \neq |\zeta_2|. \tag{4.14}$$

This condition is equivalent with the requirement that our triangle T is not

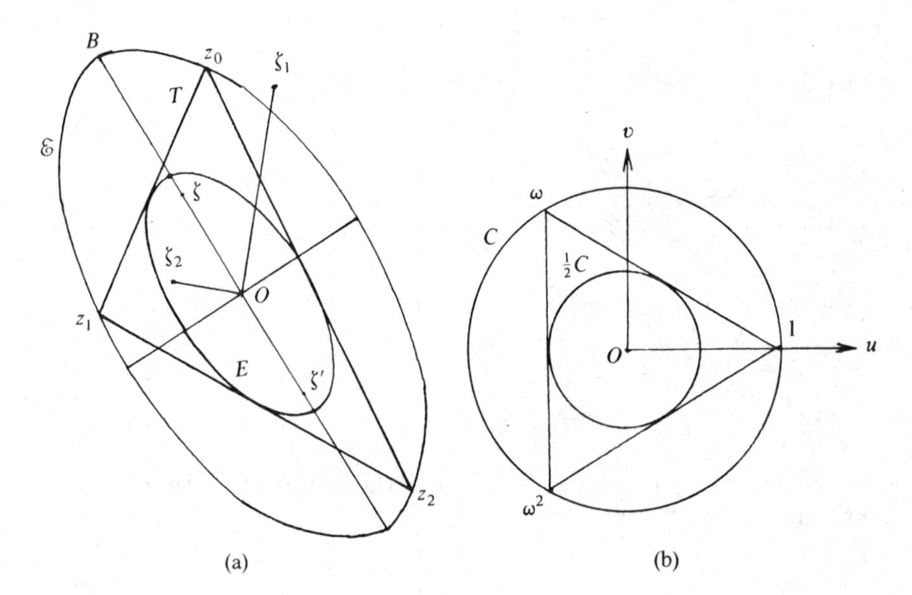

FIG. 7.3

degenerate. The case where the points z_0, z_1, z_2 are collinear will be dealt with in §5.

The role of the Steiner ellipse in our problem is the subject of

THEOREM 1 (van den Berg). *If*

$$f(z) = (z-z_0)(z-z_1)(z-z_2) = z^3 - a_1 z^2 + a_2 z - a_3 \quad (4.15)$$

is a cubic polynomial whose zeros are the vertices of the nondegenerate triangle

$$T = (z_0, z_1, z_2), \quad (4.16)$$

then the zeros of its derivative

$$f'(z) = 3z^2 - 2a_1 z + a_2 \quad (4.17)$$

coincide with the foci ζ and ζ' of the Steiner ellipse E of the triangle T.

Proof: Let a and b be the semiaxes of the ellipse (4.7). Evidently

$$a = \max|\zeta_1 e^{i\theta} + \zeta_2 e^{-i\theta}| = |\zeta_1| + |\zeta_2| \quad (4.18)$$

and

$$b = \min|\zeta_1 e^{i\theta} + \zeta_2 e^{-i\theta}| = \||\zeta_1| - |\zeta_2|\|. \quad (4.19)$$

It follows that

$$c^2 = a^2 - b^2 = (|\zeta_1| + |\zeta_2|)^2 - (|\zeta_1| - |\zeta_2|)^2 = 4|\zeta_1\zeta_2|. \quad (4.20)$$

From (4.7) and Figure 7.3(a) it is clear that the foci of $\mathfrak{E}$ are located on the bisectrix OB of $\angle\zeta_1 O\zeta_2$, at a distance c from O. By (4.20) this distance is $c = 2\sqrt{|\zeta_1\zeta_2|}$. By (4.13) the foci of E are on the same bisectrix OB, of Figure 7.3 (a), at the distance $\sqrt{|\zeta_1\zeta_2|}$ from O. This shows that the foci ζ, ζ', of E, are the roots of the quadratic equation

$$z^2 - \zeta_1\zeta_2 = 0. \quad (4.21)$$

On the other hand, the relation (3.2), or

$$3\zeta_1\zeta_2 = -(z_0 z_1 + z_0 z_2 + z_1 z_2),$$

may be written in terms of (4.15) as $3\zeta_1\zeta_2 = -a_2$. The equation (4.21) now becomes $3z^2 + a_2 = 0$, and this coincides with the quadratic $f'(z) = 0$ of (4.17), because $a_1 = 0$ by our assumptions.

5. What if the Points z_0, z_1, z_2 Are Collinear? Here we are in the case, excluded by our assumption (4.14), when

$$|\zeta_1| = |\zeta_2|. \tag{5.1}$$

Without loss of generality we may assume that $z_0 = x_0$, $z_1 = x_1$, $z_2 = x_2$ are real, for instance $x_0 < x_1 < x_2$, so that we have the degenerate triangle

$$T = (x_0, x_1, x_2).$$

Problem. Let

$$f(x) = (x - x_0)(x - x_1)(x - x_2). \tag{5.2}$$

We are to find the zeros ζ, ζ' of $f'(x) = 0$.

A first solution is given by our discussion of §4. Referring to Figure 7.2 and recalling that $z_\nu = x_\nu$ are real, and choosing the origin at their centroid so that $\zeta_0 = 0$, we construct the points ζ_1 and ζ_2 by using the equations (2.4). Therefore

$$\zeta_1 = \tfrac{1}{3}(x_0 + x_1\omega^2 + x_2\omega), \quad \zeta_2 = \tfrac{1}{3}(x_0 + x_1\omega + x_2\omega^2). \tag{5.3}$$

The location of these points in Figure 7.2 is correct, but their construction from the equations (5.3) is not indicated in Figure 7.2. Notice that ζ_1 and ζ_2 are conjugates, hence satisfy (5.1). Again we have the quadratic (4.21) for the zeros of $f'(x)$. In our case (4.21) becomes $z^2 - \zeta_1\bar{\zeta}_1 = 0$, or

$$z^2 = |\zeta_1|^2.$$

Its roots $\zeta = -|\zeta_1|$ and $\zeta' = |\zeta_1|$ are shown in Figure 7.2. The segment $[\zeta, \zeta']$ represents the degenerate Steiner ellipse of the "triangle" (5.1).

Different, perhaps more elegant, solutions of our problem are obtained by using the versatility of affine maps. We no longer need to assume that $x_0 + x_1 + x_2 = 0$. Through the origin O we draw on a slanting line $O\eta$, as shown in Figure 7.4. In the oblique coordinate system $xO\eta$ we consider the one-parameter family of affine mappings

$$A_t: \begin{array}{l} x_t = x \\ \eta_t = t\eta \quad (0 \leqslant t \leqslant 1) \end{array} \tag{5.4}$$

depending on the parameter t.

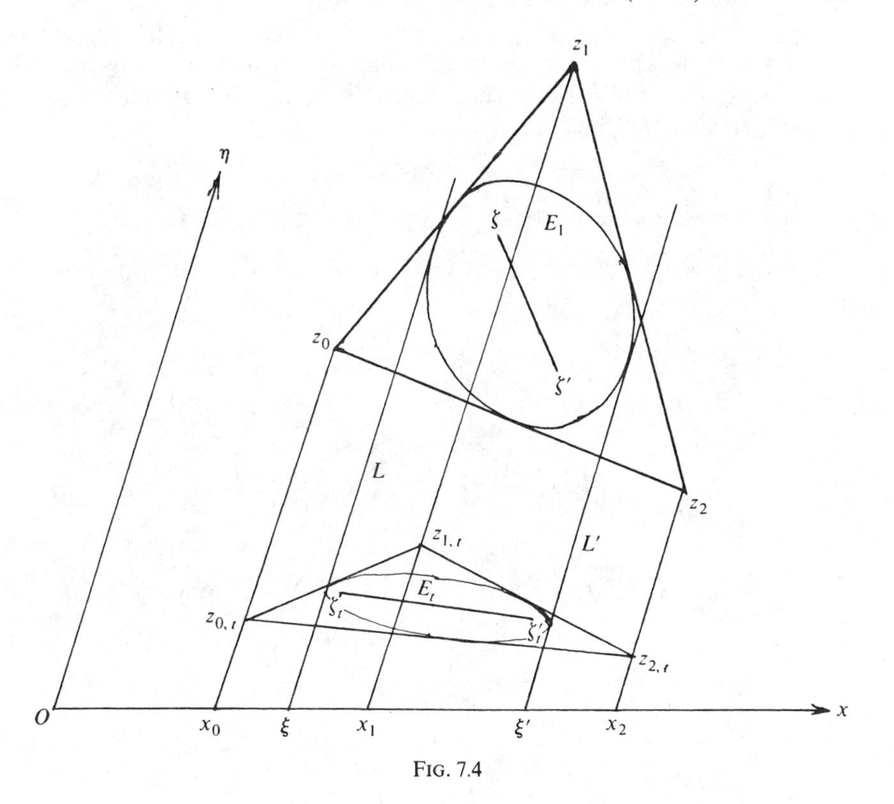

FIG. 7.4

We draw through the points x_0, x_1, x_2, parallels to $O\eta$, and pick on them the points z_0, z_1, z_2, respectively, such that the triangle

$$T_1 = (z_0, z_1, z_2) \tag{5.5}$$

be nondegenerate. Construct for T_1 its Steiner ellipse E_1 having the foci ζ, ζ'. We apply to T_1 the mapping (5.4), with $0<t<1$, obtaining the triangle

$$T_t = (z_{0,t}, z_{1,t}, z_{2,t}), \tag{5.6}$$

and let E_t be its Steiner ellipse. The crucial remark is this: Because

$$T_t = A_t T_1, \tag{5.7}$$

it follows that

$$E_t = A_t E_1,$$

and this implies that *E_t is tangent to the two tangents L and L' to E_1 which are parallel to $O\eta$. This holds for all t in $0<t\leq 1$.*

As $t \to +O$, the ellipse E_t gets flattened out, with the result that its foci ζ_t, ζ'_t, approach the endpoints of its major axis. If ξ and ξ' are the intersections of L and L' with Ox, then

$$\xi = \lim \zeta_t, \quad \xi' = \lim \zeta'_t \quad \text{as } t \to 0. \tag{5.8}$$

In the implementations of this construction we use the fact that E_1 is readily obtained *if T_1 is a regular triangle*, for in this case E_1 *is the circle*

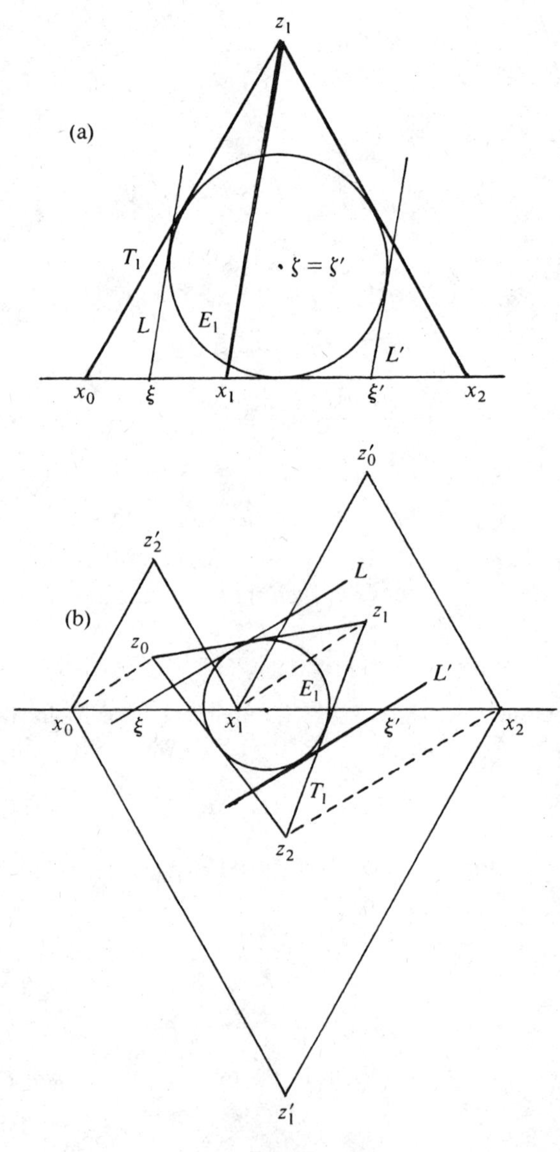

FIG. 7.5

inscribed in T_1. We conclude with two different constructions based on the general idea of Figure 7.4.

The first construction (Figure 7.5(a)). Let $T_1=(x_0, x_2, z_1)$ be the regular triangle with base x_0x_2, and let E_1 be its inscribed circle. Join z_1 to x_1 and let L and L' be the tangents to E_1 which are parallel to z_1x_1. The intersections of L and L' with Ox are the zeros of $f'(x)$ we are looking for.

The second construction (Figure 7.5 (b)). To the degenerate triangle

$$T=(x_0, x_1, x_2)$$

we apply Napoleon's theorem as stated in Problem 3 at the end of this chapter. This means that we draw the three regular triangles

$$x_0x_1z_2', \quad x_1x_2z_0', \quad x_2x_0z_1'.$$

The statement of that theorem is that the triangle

$$T_1=(z_0, z_1, z_2),$$

having their centers as vertices, is a regular triangle. Notice that the three lines

$$x_0z_0, \quad x_1z_1, \quad x_2z_2 \tag{5.9}$$

form 30° angles with Ox and are therefore parallel. Let E_1 be the circle inscribed in T_1. If we draw the tangents L and L' to E_1 which are parallel to the lines (5.9), then their intersections with the x-axis are the zeros ξ and ξ' of the quadratic $f'(x)=0$. See the paper [**4**].

Problems

1. Let $T=(z_0, z_1, z_2)$ be a triangle in the complex plane. Show that T is a regular triangle if and only if

$$z_0^2+z_1^2+z_2^2-z_0z_1-z_0z_2-z_1z_2=0.$$

2. The problem is to divide the volume of a hemisphere in two equal parts by a plane parallel to its circular base.

Hint: Assume radius of the hemisphere to be unity. Show that the distance x of the dissecting plane to the base satisfies the equation

$$x^3-3x+1=0.$$

Apply Cardan's formulae to show that $x=2\cos 80°=.3472964$. Now we know how to serve exactly one-half cup of tea in a hemispherical cup.

3. Let $T=(z_0, z_1, z_2)$ be a plane triangle, and let

$$z_1 z_0' z_2, \quad z_2 z_1' z_0, \quad z_0 z_2' z_1,$$

be the three regular triangles drawn on the sides of T toward its outside. Call t_0, t_1, t_2 the centers of these triangles. Establish

NAPOLEON'S THEOREM. *The triangle* $t=(t_0, t_1, t_2)$ *is regular.*

Hint: Express z_0', z_1', z_2' in terms of z_0, z_1, z_2. From these derive expressions for the t_ν, and finally verify that $(t_2-t_1)\omega=t_0-t_2$, where $\omega=\exp(2\pi i/3)$. See Honsberger's book [**2**, pp. 35–36] for a different proof.

References

[**1**] F. J. van den Berg, *Nogmals over afgeleide Wortelpunten*, Nieuw Arch. Wisk., 15 (1888) 100–164.

[**2**] Ross Honsberger, *Mathematical Gems*, Mathematical Association of America, 1973.

[**3**] Morris Marden, *Geometry of polynomials*, Amer. Math. Soc., Math. Surveys, No. 3, 1966.

[**4**] I. J. Schoenberg, *On two theorems of Archimedes and F. J. van den Berg*, Simon Stevin, 35 (1962) 133–138.

[**5**] D. J. Struik, *A Concise History of Mathematics*, 3rd ed., Dover Publications, New York, 1967.

[**6**] J. L. Walsh, *On the location of the roots of the derivative of a polynomial*, Ann. of Math., 22 (1920) 128–144.

CHAPTER 8

THE FINITE FOURIER SERIES III: J. STEINER AND L. FEJES TÓTH

1. Introduction. We return for a moment to Kasner's theorem of Chapter 6, §1A, to introduce a useful term. If

$$\Pi = (z_0, z_1, \ldots, z_{k-1}) \tag{1.1}$$

is a polygon and

$$\Pi' = (z_0', z_1', \ldots, z_{k-1}') \tag{1.2}$$

is the polygon having as vertices the midpoints $z_\nu' = \frac{1}{2}(z_\nu + z_{\nu+1})$ of the sides of Π, then we say that Π' is the *Kasner polygon of* Π and write

$$\Pi' = K\Pi. \tag{1.3}$$

If we think of Π' as given, then we say that Π is the *pre-Kasner polygon of* Π', writing

$$\Pi = K^{-1}\Pi'. \tag{1.4}$$

However, we know from Kasner's theorem that Π need not exist if k is even. For this reason, *we shall consider pre-Kasner polygons only if*

$$k \text{ is odd} \tag{1.5}$$

when they exist uniquely.

Several problems present themselves if we iterate the formation of the Kasner polygons. For instance, we may repeat the operation (1.3) obtaining a sequence of k-gons $\Pi' = K\Pi$, $\Pi'' = K\Pi' = K^2\Pi, \ldots,$ and generally

$$\Pi^{(n)} = K^{(n)}\Pi \qquad (n=1,2,\ldots) \text{ (Figure 8.1)}. \tag{1.6}$$

What can we say about the nth-order Kasner polygon $\Pi^{(n)}$ as $n \to \infty$?

Here is another question: Assume that k is odd and that we form the pre-Kasner polygons $\Pi^{(-1)} = K^{-1}\Pi$, $\Pi^{(-2)} = K^{-1}\Pi^{(-1)} = K^{-2}\Pi, \ldots,$ and

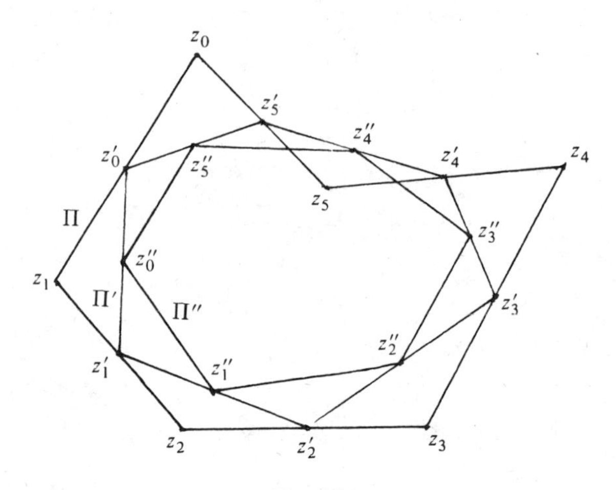

FIG. 8.1

generally

$$\Pi^{(-n)} = K^{-n}\Pi \qquad (n=1,2,\ldots). \tag{1.7}$$

How does $\Pi^{(-n)}$ behave as $n \to \infty$?

A second topic that we study in the present MTE is a classical result of Jakob Steiner.

THEOREM OF STEINER. *Among all k-gons in the plane having the same perimeter L, the k-gon having largest area is the regular k-gon.*

The competing k-gons are said to be *isoperimetric*, and the theorem is called, not very logically, the *isoperimetric property* of the regular k-gon.

In dealing with this theorem (§2) we will assume that all competing k-gons Π are not only isoperimetric but also *equilateral*; hence having all their sides $=L/k$. But this is no real restriction: If Π is not equilateral, then some two adjacent sides are unequal, say

$$|z_1 - z_0| < |z_2 - z_1|.$$

I claim that Π cannot have maximal area among all k-gons of perimeter L: Let E be the ellipse of foci z_0 and z_2 passing through z_1 (Figure 8.2), and let z_1^* be the vertex of E on its minor axis. The two polygons

$$\Pi = (z_0, z_1, z_2, \ldots, z_{k-1}) \quad \text{and} \quad \Pi^* = (z_0, z_1^*, z_2, \ldots, z_{k-1})$$

have the same perimeter L, while Π^* has visibly larger area.

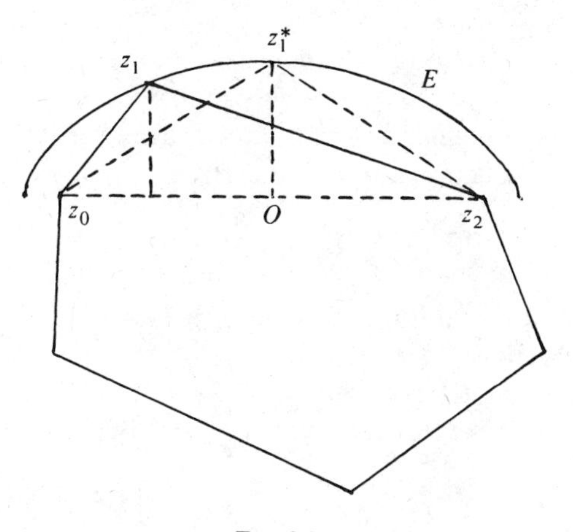

FIG. 8.2

The topics of this Introduction were chosen to show the power and flexibility of the finite Fourier series.

2. The Isoperimetric Property of the Regular k-gon. Let

$$\Pi = (z_0, z_1, \ldots, z_{k-1}) \tag{2.1}$$

be an equilateral k-gon, and let

$$z_\alpha = \sum_{\nu=0}^{k-1} \zeta_\nu \omega_\alpha^\nu \qquad (\alpha = 0, \ldots, k-1) \tag{2.2}$$

be the f.F.s. of its vertices. Furthermore, let

$$L = \text{Perimeter of } \Pi, \quad A = \text{Area of } \Pi. \tag{2.3}$$

Our proof of Steiner's theorem is based on the following remarkable *identity*:

$$L^2 - \left(4k \tan\frac{\pi}{k}\right)A = \sum_{\nu=2}^{k-1} 4k^2 \sin\frac{\pi\nu}{k}\left(\sin\frac{\pi\nu}{k} - \tan\frac{\pi}{k}\cos\frac{\pi\nu}{k}\right) \cdot |\zeta_\nu|^2. \tag{2.4}$$

Let us use the abbreviation

$$I_k(L, A) = L^2 - \left(4k \tan\frac{\pi}{k}\right)A. \tag{2.5}$$

Notice the following:

1. The terms for $\nu=0$ and $\nu=1$ are absent on the right-hand side of (2.4).
2. The coefficient of $|\zeta_\nu|^2$ is >0 for $\nu=2,3,\dots,k-1$.
3. If $\Pi=\Pi_r$ is the regular k-gon, then we easily verify that its area A_r is given by $A_r=L^2/(4k\tan(\pi/k))$, and therefore

$$I_k(L,A_r)=0. \tag{2.6}$$

However, this result also follows from (2.4): If $\Pi=\Pi_r$, then $\zeta_2=\zeta_3=\cdots=\zeta_{k-1}=0$ and (2.4) shows that (2.6) holds.

If Π is *not* the regular k-gon, then in (2.2) at least one of the coefficients $\zeta_2,\dots,\zeta_{k-1}$ must be different from zero, and so, by (2.4),

$$L^2-\left(4k\tan\frac{\pi}{k}\right)A>0$$

and therefore

$$A<L^2/\left(4k\tan\frac{\pi}{k}\right)=A_r. \tag{2.7}$$

This establishes Steiner's theorem.

Proof of the identity (2.4). This is done with the f.F.s. in two stages, the first expressing the perimeter L in terms of the ζ_ν, the second doing the same for the area A. In both stages we assume Π to be equilateral; hence

$$|z_\nu-z_{\nu+1}|=L/k=a \qquad (\nu=0,\dots,k-1,\ z_k=z_0). \tag{2.8}$$

(i) *Expressing L in terms of the* ζ_ν. From the sequence z_ν we form the new sequence z'_ν by the equations

$$z'_\nu=z_\nu-z_{\nu+1} \qquad (\nu=0,\dots,k-1), \quad \text{where } z_k=z_0. \tag{2.9}$$

Notice that this is a linear transformation which is precisely of the cyclic type (5.6) of Theorem 2 of Chapter 6, its representative polynomial having the simple form

$$f(z)=-1+z,$$

for (2.9) becomes identical with (5.6) if we choose $a_0=-1$, $a_1=1$, $a_2=\cdots=a_{k-1}=0$. But then the corollary (5.9) in Chapter 6 shows that we have the identity

$$\sum_0^{k-1}|z_\nu-z_{\nu+1}|^2=k\sum_0^{k-1}|\omega_\nu-1|^2|\zeta_\nu|^2. \tag{2.10}$$

By (2.8) the left-hand side is equal to $k(L/k)^2 = L^2/k$. Since $|\omega_\nu - 1| = 2\sin(\pi\nu/k)$, we may rewrite (2.10) as

$$L^2 = \sum_{\nu=1}^{k-1} 4k^2 \sin^2 \frac{\pi\nu}{k} \cdot |\zeta_\nu|^2. \tag{2.11}$$

This is the expression of L^2 that we need.

(ii) *Expressing A in terms of the ζ_ν.* We assume the reader to know that the area A of the polygon (1.1) is given by the relation

$$A = \sum_{\nu=0}^{k-1} (\text{Area of triangle } Oz_\nu z_{\nu+1}) \tag{2.12}$$

where we use the positive or negative areas of oriented polygons and triangles. To bridge the gap between A and the complex f.F.s it is clear that we must use the real and imaginary parts of $z_\nu = x_\nu + iy_\nu$. From analytic geometry we use the familiar formula

$$\text{area of triangle } Oz_\nu z_{\nu+1} = \tfrac{1}{2}(x_\nu y_{\nu+1} - x_{\nu+1} y_\nu). \tag{2.13}$$

However

$$\begin{aligned} x_\nu y_{\nu+1} - x_{\nu+1} y_\nu &= \operatorname{Im}\{(x_\nu - iy_\nu)(x_{\nu+1} + iy_{\nu+1})\} \\ &= \operatorname{Im}(\bar{z}_\nu z_{\nu+1}), \end{aligned}$$

and so

$$A = \frac{1}{2} \sum_{\nu=0}^{k-1} \operatorname{Im}(\bar{z}_\nu z_{\nu+1}) = \frac{1}{2} \operatorname{Im} \sum_{\nu=0}^{k-1} \bar{z}_\nu z_{\nu+1}. \tag{2.14}$$

On the right we recognize a convolution of sequences, and to evaluate it we use the f.F.s. (2.2). We have

$$\begin{aligned} \sum_{\nu=0}^{k-1} \bar{z}_\nu z_{\nu+1} &= \sum_\nu \Big(\sum_\alpha \bar{\zeta}_\alpha \bar{\omega}_\nu^\alpha \Big) \Big(\sum_\beta \zeta_\beta \omega_{\nu+1}^\beta \Big) = \sum_{\alpha,\beta,\nu} \bar{\zeta}_\alpha \zeta_\beta \bar{\omega}_\nu^\alpha \omega_{\nu+1}^\beta \\ &= \sum_{\alpha,\beta} \bar{\zeta}_\alpha \zeta_\beta \omega_1^\beta \sum_{\nu=0}^{k-1} \bar{\omega}_\nu^\alpha \omega_\nu^\beta, \end{aligned}$$

and the orthogonality relations (1.21) in Chapter 6 simplify this to

$$\sum_{\nu=0}^{k-1} \bar{z}_\nu z_{\nu+1} = k \sum_{\alpha=0}^{k-1} |\zeta_\alpha|^2 \omega_\alpha.$$

Taking on both sides of the imaginary parts and using (2.14) we finally obtain that

$$A=\frac{k}{2}\sum_{\alpha=1}^{k-1}|\zeta_\alpha|^2\sin\frac{2\pi\alpha}{k}. \tag{2.15}$$

To obtain (2.4) we merely need to combine the results (2.11) and (2.15). We multiply (2.15) by $4k\tan(\pi/k)$ to obtain

$$\left(4k\tan\frac{\pi}{k}\right)A=\sum_{\nu=1}^{k-1}4k^2\tan\frac{\pi}{k}\sin\frac{\pi\nu}{k}\cos\frac{\pi\nu}{k}\cdot|\zeta_\nu|^2, \tag{2.16}$$

and observe that here and also in (2.11) the coefficients of $|\zeta_1|^2$ are both equal to $4k^2\sin^2(\pi/k)$. On subtracting (2.16) from (2.11) these equal terms cancel and we obtain

$$L^2-\left(4k\tan\frac{\pi}{k}\right)A=\sum_{\nu=2}^{k-1}4k^2\sin\frac{\pi\nu}{k}\left(\sin\frac{\pi\nu}{k}-\tan\frac{\pi}{k}\cos\frac{\pi\nu}{k}\right)|\zeta_\nu|^2,$$

and the identity (2.4) is established.

REMARKS. 1. My guess is that Jakob Steiner, who was a great geometer averse to algebraic calculations, would have heartily disliked our proof of his theorem. A somewhat longer version of our proof, avoiding entirely the use of complex numbers, is due to W. Blaschke. For references see [**1**, Reference 7 at the end].

2. If in the inequality

$$A\leqslant L^2/\left(4k\tan\frac{\pi}{k}\right)$$

we let $k\to\infty$, we obtain the inequality

$$A\leqslant\frac{1}{4\pi}L^2. \tag{2.17}$$

This is the famous *isoperimetric inequality* for closed curves in the plane. The equality sign holds only for circles. (2.17) can be established by Fourier series. For references see again [**1**, Reference 7].

3. The Behavior of the Kasner Polygons $\Pi^{(n)}$ as $n\to\infty$. A Theorem of L. Fejes Tóth. Let

$$\Pi=(z_0,z_1,\ldots,z_{k-1}) \tag{3.1}$$

be a k-gon and let

$$\Pi^{(n)} = K^n\Pi = \left(z_0^{(n)}, z_1^{(n)}, \ldots, z_{k-1}^{(n)}\right) \tag{3.2}$$

be its nth Kasner polygon (see §1 for its definition).

THEOREM OF L. FEJES TÓTH. *An appropriate affine image of $\Pi^{(n)}$ converges to a regular k-gon, or else to a star-shaped regular k-gon, as $n \to \infty$.*

This theorem seems tailor made for an application of the f.F.s., especially of Theorem 2 in Chapter 6, §5. Indeed, we get the vertices of

$$K\Pi = \left(z_0', \ldots, z_{k-1}'\right)$$

from

$$z_\nu' = \tfrac{1}{2}(z_\nu + z_{\nu+1}) \qquad (\nu = 0, \ldots, k-1;\ z_k = z_0), \tag{3.3}$$

and this is the *cyclic transformation* (5.6) in Chapter 6 for the special representative polynomial

$$f(z) = \tfrac{1}{2}(1+z). \tag{3.4}$$

By Theorem 2, in particular the relations (5.7) in Chapter 6, we conclude that the f.F. coefficients ζ_ν' of $\Pi' = K\Pi$ are

$$\zeta_\nu' = \zeta_\nu \frac{1+\omega_\nu}{2} \qquad (\nu = 0, \ldots, k-1). \tag{3.5}$$

This result may now be iterated and shows that the f.F. coefficients of $\Pi^{(n)}$ are

$$\zeta_\nu^{(n)} = \zeta_\nu \left(\frac{1+\omega_\nu}{2}\right)^n \qquad (\nu = 0, \ldots, k-1). \tag{3.6}$$

Passing to the vertices of the polygon (3.2) we find that

$$z_\nu^{(n)} = \zeta_1\left(\frac{1+\omega_1}{2}\right)^n \omega_\nu + \zeta_2\left(\frac{1+\omega_2}{2}\right)^n \omega_\nu^2 + \cdots + \zeta_{k-1}\left(\frac{1+\omega_{k-1}}{2}\right)^n \omega_\nu^{k-1}, \tag{3.7}$$

where we have assumed that the centroid of the vertices of Π is at 0; hence $\zeta_0 = 0$. Here it is important to observe that

$$\frac{1+\omega_\nu}{2} = \cos\frac{\pi\nu}{k} e^{i\pi\nu/k} = \cos\frac{\pi\nu}{k} \cdot \alpha^\nu \tag{3.8}$$

(see Figure 8.3) where $\alpha = e^{i\pi/k}$. Also that their absolute values

$$\left|\frac{1+\omega_\nu}{2}\right| = \left|\cos\frac{\pi\nu}{k}\right| \qquad (\nu = 1,\ldots,k-1)$$

form a "unimodal" sequence symmetric about its middle; also that the extreme terms (for $\nu = 1$, $\nu = k-1$) are equal and are the largest of the sequence, the terms decreasing as we approach the middle. This implies that for very large values of n only the two extreme terms of the sum (3.7) will determine the shape of the polygon $\Pi^{(n)}$ because they are the *leading* terms for large n. The sum of these leading terms is

$$\zeta_1\left(\cos\frac{\pi}{k}\right)^n \alpha^n\omega_\nu + \zeta_{k-1}\left(-\cos\frac{\pi}{k}\right)^n(-\bar{\alpha})^n\omega_\nu^{k-1}$$
$$= \zeta_1\left(\cos\frac{\pi}{k}\right)^n \alpha^n\omega_\nu + \zeta_{-1}\left(\cos\frac{\pi}{k}\right)^n \bar{\alpha}^n\bar{\omega}_\nu.$$

Here we have written $\omega = \omega_1 = \exp(2\pi i/k)$.

Let us use these facts to determine the shape of $\Pi^{(n)}$ for large values of n. We assume that

$$|\zeta_1| + |\zeta_{-1}| > 0, \tag{3.9}$$

and write

$$\gamma = \cos\frac{2\pi}{k}\Big/\cos\frac{\pi}{k} \qquad (0<\gamma<1 \text{ if } k>4). \tag{3.10}$$

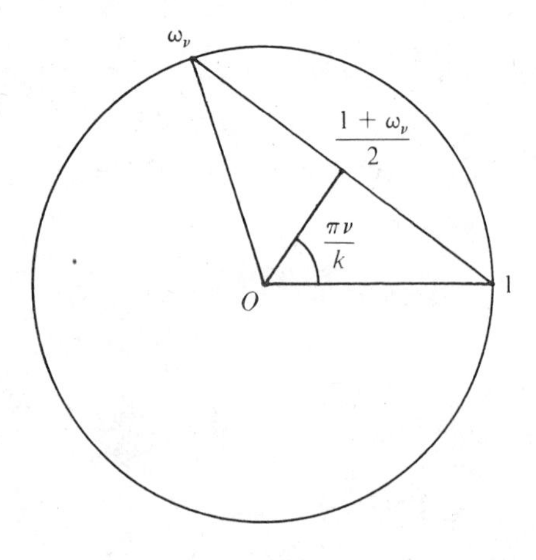

FIG. 8.3

If we divide the relations (3.7) by $\left(\cos\frac{\pi}{k}\right)^n$ we obtain

$$z_\nu^{(n)}\Big/\left(\cos\frac{\pi}{k}\right)^n = \zeta_1\alpha^n\omega_\nu + \zeta_{-1}\bar{\alpha}^n\bar{\omega}_\nu + 0(\gamma^n) \tag{3.11}$$

where all terms of (3.7), except the two extreme ones, have been absorbed in the error term.

We see from (3.11) that, as $n\to\infty$, $\Pi^{(n)}$ becomes similar to the affine regular k-gons

$$\zeta_1\alpha^n\omega_\nu + \zeta_{-1}\bar{\alpha}^n\bar{\omega}_\nu \qquad (\nu=0,\ldots,k-1). \tag{3.12}$$

Notice that increasing n by $2k$ will not change (3.12). There are therefore at most $2k$ limiting shapes.

If (3.9) does not hold, while $\zeta_1=\zeta_{-1}=0$ and $|\zeta_2|+|\zeta_{-2}|>0$, say, then a similar argument will show that there are affine regular *star-shaped* k-gons as limiting shapes of $\Pi^{(n)}$.

This completes our proof of L. Fejes Tóth's theorem. See [2] also for further references.

REMARKS. 1. Fritz John tells me that he conjectured Fejes Tóth's theorem while still in high school and actually proved it for $k=5$ by elementary geometry.

2. Properties of the kind described here are usually called *asymptotic properties*. The name comes from the hyperbola and its asymptotes.

4. What about the Pre-Kasner Polygons $\Pi^{(-n)}$ as $n\to+\infty$? This is a variation of the previous problem, but requires that

$$k \text{ is odd}, \quad k = 2m+1. \tag{4.1}$$

We assume this and know that all $\Pi^{(-n)}$ exist uniquely (see §1). In (3.2) we replace n by $-n$ to obtain

$$\Pi^{(-n)} = K^{-n}\Pi = \left(z_0^{(-n)},\ldots,z_{k-1}^{(-n)}\right). \tag{4.2}$$

Since Π is the Kasner polygon of $\Pi^{(-1)}$, we obtain from (3.5) the relations

$$\zeta_\nu = \zeta_\nu^{(-1)}\frac{1+\omega_\nu}{2},$$

and therefore

$$\zeta_\nu^{(-1)} = \zeta_\nu\left(\frac{1+\omega_\nu}{2}\right)^{-1} \qquad (\nu=0,\ldots,k-1).$$

Iterating this result we gather that

$$\zeta_\nu^{(-n)} = \zeta_\nu \left(\frac{1+\omega_\nu}{2} \right)^{-n} \qquad (n=1,2,\ldots). \tag{4.3}$$

Assuming that $\zeta_0 = 0$, hence $\zeta_0^{(-n)} = 0$, we find, by (4.3), for the f.F.s. of (4.2), instead of (3.7), the relations

$$z_\nu^{(-n)} = \zeta_1 \left(\frac{1+\omega_1}{2} \right)^{-n} \omega_\nu + \zeta_2 \left(\frac{1+\omega_2}{2} \right)^{-n} \omega_\nu^2$$

$$+ \cdots + \zeta_{k-1} \left(\frac{1+\omega_{k-1}}{2} \right)^{-n} \omega_\nu^{k-1}. \tag{4.4}$$

It should be clear that $\Pi^{(-n)}$ will likely become large as n increases; to recognize its shape we must shrink $\Pi^{(-n)}$ to manageable size, and (4.4) shows how to do that. Again the sequence (3.8) decides the outcome. However, since (4.4) depends on the nth powers of the *reciprocals* of (3.8), this time the two central terms of (4.4) are the leading terms of the right-hand side of (4.4). Fortunately, by (4.1), there are two central terms corresponding to

$$\nu = m = \frac{k-1}{2} \quad \text{and} \quad \nu = m+1 = \frac{k+1}{2}.$$

To ensure that these are really the leading terms, we must assume that

$$|\zeta_m| + |\zeta_{m+1}| > 0. \tag{4.5}$$

Careful evaluation of these central terms of (4.4) shows the following:
If we multiply (4.4) by $(\sin(\pi/2k))^n$, then (4.4) reduces to

$$z_\nu^{(-n)} \left(\sin \frac{\pi}{2k} \right)^n = \zeta_m i^n e^{(i\pi n/2k)} \omega_m^\nu$$

$$+ \zeta_{m+1} (-i)^n e^{-i\pi n/2k} \overline{\omega}_m^\nu + 0(\gamma_1^n) \quad \text{as } n \to \infty \tag{4.6}$$

where

$$\gamma_1 = \sin \frac{\pi}{2k} \Big/ \sin \frac{3\pi}{2k}. \tag{4.7}$$

For a fixed but large value of n we see that the right-hand side of (4.6) is uniformly close to

$$\tilde{z}_\nu^{(-n)} = \zeta_m (a_k)^n \omega_m^\nu + \zeta_{m+1} (\overline{a}_k)^n \overline{\omega}_m^\nu \qquad (\nu = 0, \ldots, k-1), \tag{4.8}$$

where $a_k = i e^{\pi i/2k}$ is a $(4k)$th root of unity. This proves the

THEOREM. *If $k=2m+1$ is odd and* (4.5) *holds, then an appropriate affine image of $\Pi^{(-n)}$ converges to the regular star-shaped k-gon*

$$e^{(2\pi i m/k)\cdot\nu} \qquad (\nu=0,\ldots,k-1) \qquad (k=2m+1). \tag{4.9}$$

Observe that (4.9) *is the kinkiest among all star-shaped k-gons.*

If $\zeta_m=\zeta_{m+1}=0$ while $|\zeta_{m-1}|+|\zeta_{m+2}|>0$, then the next (less kinky) star-shaped k-gon will take over as limiting shape. It does seem remarkable that, if $k=5$, then the limiting shape of $\Pi^{(-n)}$ is a *star-shaped pentagon*, at least in the general case.

Problems

1. Use Theorem 2 of Chapter 6, §5, to show that for a k-gon $\Pi=(z_\nu)$ and its Kasner k-gon $\Pi'=(z'_\nu)$, the identity

$$\sum_0^{k-1} |z'_\nu|^2 = k\sum_{\nu=0}^{k-1} \cos^2\frac{\pi\nu}{k}\cdot|\zeta_\nu|^2$$

holds. Conclude from this and the Parseval relation (4.3) of Chapter 6, §4, that the nth Kasner polygon $\Pi^{(n)}$ shrinks to the center of gravity of the vertices of Π, as $n\to\infty$.

2. Let $\Pi=P_0P_1\ldots P_{k-1}$ be a closed plane k-gon having the point O as centroid of its vertices. Use the identity (2.11) and the Parseval relation (4.3) of Chapter 6, §4 to prove the following two theorems:

(i) Let $k=4$ so that Π is a quadrilateral. Prove the inequalities

$$2\leqslant\frac{\sum_0^3 (P_\nu P_{\nu+1})^2}{\sum_0^3 (OP_\nu)^2}\leqslant 4$$

with equality on the left if and only if $\Pi=P_0P_1P_2P_3$ is a parallelogram, and equality on the right if and only if $P_0=P_2$ and $P_1=P_3$.

(ii) Let $k=5$, so that $\Pi=P_0P_1P_2P_3P_4$ is a pentagon. Prove that

$$4\sin^2\frac{\pi}{5}\leqslant\frac{\sum_0^4 (P_\nu P_{\nu+1})^2}{\sum_0^4 (OP_\nu)^2}\leqslant 4\sin^2\frac{2\pi}{5},$$

with equality on the left if and only if Π is an affine regular pentagon, and

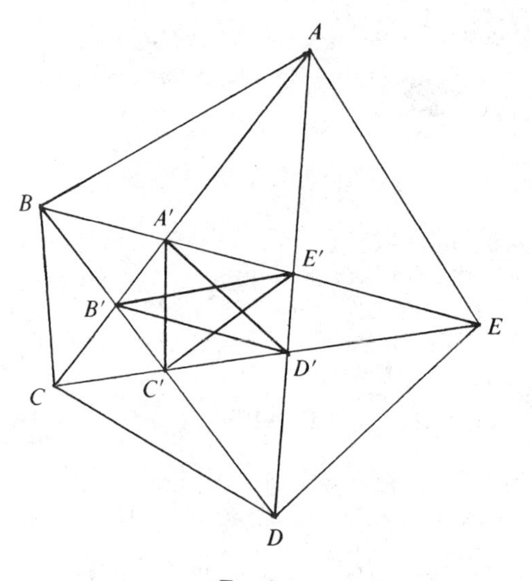

FIG. 8.4

equality on the right if and only if Π is an affine regular star-shaped pentagon.

3. I wish that I could propose the following as a problem, but I can't because I cannot do it myself. We must therefore call it a conjecture.

CONJECTURE. Let $\Pi = ABCDE$ be a strictly convex pentagon, where "strictly" means that no three vertices are collinear. We apply a "peeling process" as follows: We remove from Π the five triangles

$$ABC, \quad BCD, \quad CDE, \quad DEA, \quad \text{and} \quad EAB,$$

and are left with the "peeled" pentagon $\Pi' = A'B'C'D'E'$ of Figure 8.4, which is evidently also strictly convex. Repeating the same peeling operation on Π' we obtain the strictly convex pentagon $\Pi'' = A''B''C''D''E''$. Let $\Pi^{(n)}$ be the pentagon obtained after n operations. Clearly all these pentagons are closed sets and we have the set inclusions

$$\Pi \supset \Pi' \supset \cdots \supset \Pi^{(n)} \supset \cdots.$$

The conjecture is that $\Pi^{(n)}$ shrinks to a point as $n \to \infty$. (See [**2**, p. 9].)

References

[**1**] I. J. Schoenberg, *The finite Fourier series and elementary geometry*, Amer. Math. Monthly, 57 (1950) 390–404.

[**2**] I. J. Schoenberg, *Remarks on two geometric conjectures of L. Fejes Tóth*, An. Stiint. Univ. "Al. I. Cuza" Jassy, XXI, s. I a, 1975, 9–13.

CHAPTER 9

THE FINITE FOURIER SERIES IV: THE HARMONIC ANALYSIS OF SKEW POLYGONS AS A SOURCE OF OUTDOOR SCULPTURES

1. Introduction. What is an outdoor sculpture? It is a figure of some size that people, not necessarily mathematicians, will get aesthetic pleasure from looking at. Of course a mathematical understanding of the construction, and of the problem that it solves, will enhance the pleasure of the beholder, just as religious feelings will magnify the appreciation of a religious work of art, whether a painting, a sculpture, or a musical composition.

We begin by stating a beautiful theorem of Jesse Douglas, of Plateau problem fame, on skew pentagons [2]. By a skew pentagon $\Pi=(z_0,z_1,z_2,z_3,z_4)$ we mean a closed pentagon whose vertices $z_0,z_1,\ldots,z_4$ are arbitrarily chosen points of our three-dimensional space R^3. We say that a pentagon $\Pi^1=(f_0^1,f_1^1,f_2^1,f_3^1,f_4^1)$ in space is *affine regular*, provided that it lies in a plane π_1, and that it is in π_1 the affine image of a regular pentagon. Likewise we say that a pentagon $\Pi^2=(f_0^2,f_1^2,f_2^2,f_3^2,f_4^2)$ is *affine star-shaped regular* if it lies in a plane π_2 and is in π_2 an affine image of a star-shaped regular pentagon. I remind the reader that affine regular polygons were mentioned repeatedly in our previous essays on the f.F.s., for instance, in Chapter 6, §1, Problem 2. The theorem of Douglas is as follows:

THEOREM 1 (J. Douglas). *Let*

$$\Pi=(z_0,z_1,z_2,z_3,z_4) \tag{1.1}$$

be a skew closed pentagon in R^3, viewed as a vector space. Let

$$z'_\nu=\tfrac{1}{2}(z_{\nu+2}+z_{\nu-2})\qquad(\nu=0,1,\ldots,4)\qquad(z_{\nu+5}=z_\nu) \tag{1.2}$$

be the midpoint of the side $[z_{\nu-2},z_{\nu+2}]$ which is opposite to the vertex z_ν.

For each ν determine, on the line joining z_ν to z'_ν, the points f_ν^1, f_ν^2, such that

$$f_\nu^1-z'_\nu=\frac{1}{\sqrt{5}}(z'_\nu-z_\nu),\quad f_\nu^2-z'_\nu=-\frac{1}{\sqrt{5}}(z'_\nu-z_\nu). \tag{1.3}$$

Then

$$\Pi^1 = \left(f_0^1, f_1^1, f_2^1, f_3^1, f_4^1\right) \tag{1.4}$$

is a plane and affine regular pentagon, and

$$\Pi^2 = \left(f_0^2, f_1^2, f_2^2, f_3^2, f_4^2\right) \tag{1.5}$$

is a plane and affine regular star-shaped pentagon.

Theorem 1 is not difficult to verify, but was not easy to discover. Douglas explores thoroughly these problems for pentagons and for polygons with more vertices. He uses eigenvalue properties of cyclic, or circulant, square matrices. Theorem 1 is established directly in [**2**], with a short ad hoc proof which does not seem particularly transparent. Here we wish to point out the following.

1. The natural foundation of Douglas's idea seems to be the finite Fourier series. To be sure, the f.F.s. is essentially equivalent to the properties of cyclic matrices used by Douglas. However, it is shown in §2 that, if we invert the f.F.s. for a pentagon, not in its usual complex form but in its so-called real form, we are inevitably led to Douglas's Theorem 1. From this point of view Douglas's idea easily generalizes to the harmonic analysis of skew heptagons in R^3 (Theorem 2 of §4).

2. It is doubtful if Douglas ever built a model illustrating his theorem. The author constructed out of twenty thin wooden sticks such a model, well over two feet in size. The appearance of the plane and affine regular pentagons Π^1 and Π^2 was expected, but enjoyable just the same, especially as they lie in two different planes. For contrast, the sides of the pentagons Π, Π^1, Π^2 were painted in three different colors. The shape of the entire structure, i.e., ignoring rigid motion and a scaling factor, is easily found to depend on eight real parameters. This diversity and total lack of symmetry allows for artistic effects and makes the presence of the affine regular pentagons more striking: Order out of chaos. Made of metal bars and of a more heroic size, it would provide a striking outdoor sculpture. Our Figure 9.1 shows the case when the pentagon Π, having vertices $z_0,\ldots,z_4$, is in a plane. This, however, gives only a faint idea of the aspect of a three-dimensional structure. The reader is urged to build one. (See photo on p. 106.)

The author also constructed a three-dimensional illustration of Theorem 2. It is much more complicated, as it requires 63 thin wooden sticks. Based on a skew heptagon Π, it shows the three affine regular heptagons Π^1, Π^2, Π^3 painted in three contrasting colors. Using dowel sticks, ($\frac{1}{8}$ of an inch in diameter) for the sides of Π, all other sticks should be of walnut wood with square cross-sections, having sides equal to $\left(\frac{1}{16}\right)''$. All instructions are contained in §5. Our figure shows an example when the heptagon $\Pi = (z_0,\ldots,z_6)$ is in the plane. (See Figure 9.3 and photo on p. 116.)

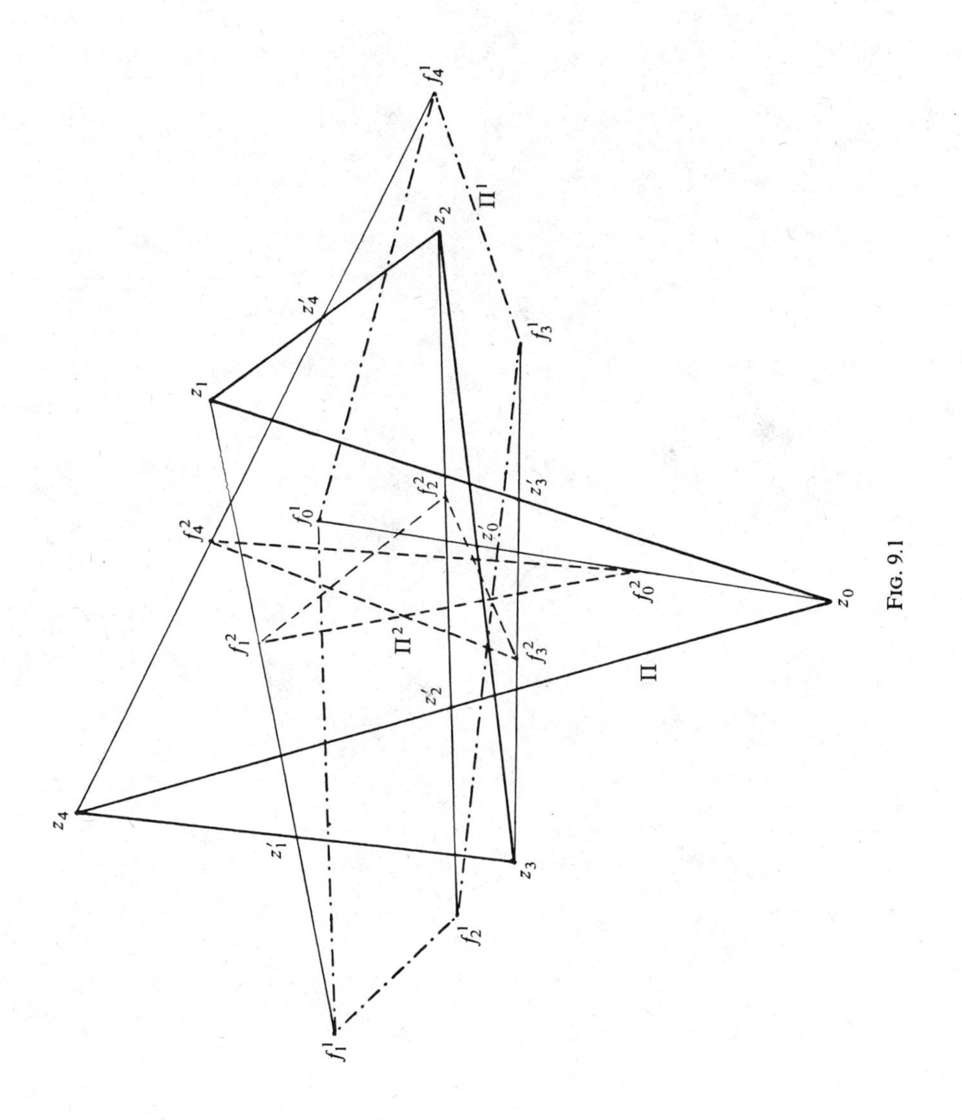

z_0
z_1
z_2
z_3
z_4
z'_0
z'_1
z'_2
z'_3
z'_4
f_0^1
f_1^1
f_2^1
f_3^1
f_4^1
f_0^2
f_1^2
f_2^2
f_3^2
f_4^2
Π
Π^1
Π^2

FIG. 9.1

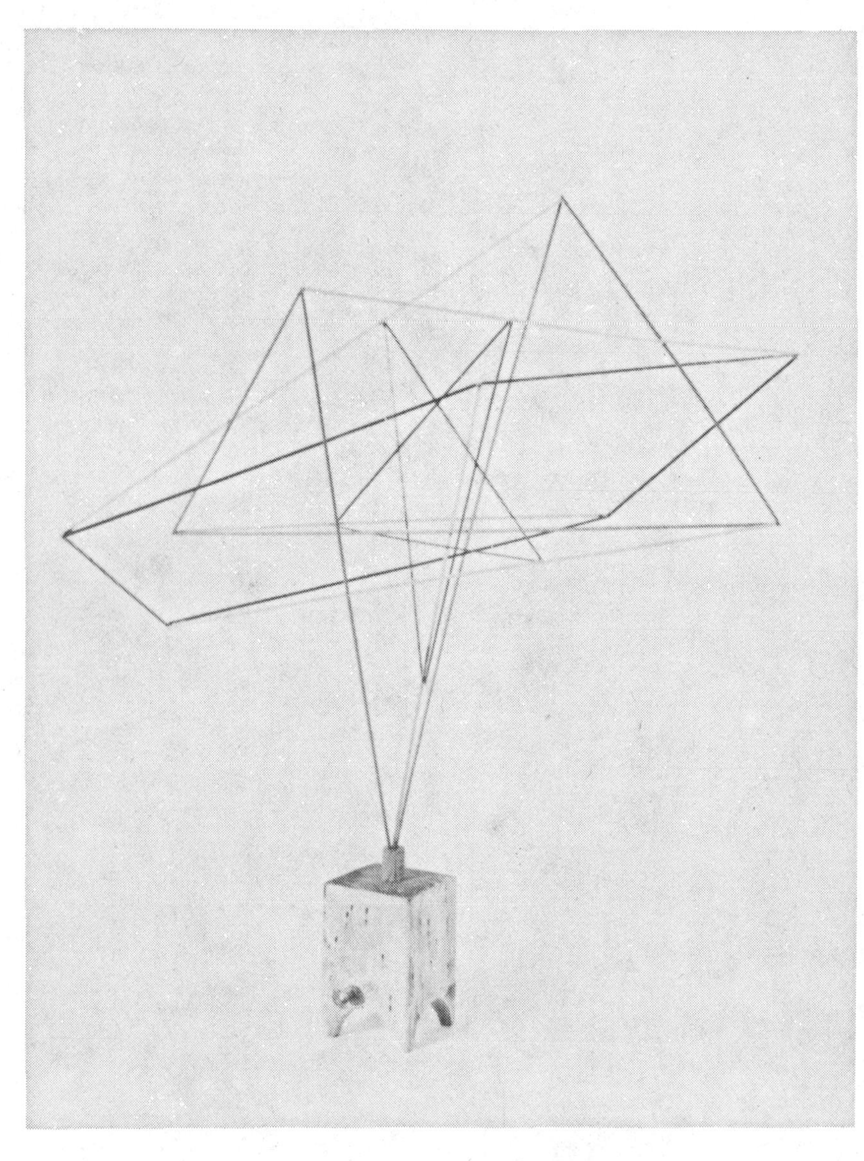

A model of the harmonic analysis of a skew pentagon (Douglas' theorem). (See Figure 9.1.)

The reader will ask: Why did we neglect to discuss the skew hexagons? The reason seems obvious, because the *star-shaped* regular hexagons are not particularly interesting, being regular triangles described twice or segments (diameters) described six times.

2. A Proof of Theorem 1 for Pentagons Π in the Complex Plane. If $\Pi \subset \mathbb{C}$ we can consider all symbols z_ν, z'_ν, f_ν^1, f_ν^2, of Theorem 1, as complex numbers. With $\omega_\nu = \exp(2\pi i\nu/5)$ the f.F.s. of the z_ν is the expansion

$$z_\nu = \zeta_0 + \zeta_1\omega_\nu + \zeta_2\omega_\nu^2 + \zeta_3\omega_\nu^3 + \zeta_4\omega_\nu^4 \qquad (\nu = 0,\dots,4) \tag{2.1}$$

where the f.F. coefficients ζ_ν are given by the inverse formulae

$$\zeta_\nu = \tfrac{1}{5}\left(z_0 + z_1\bar{\omega}_\nu + z_2\bar{\omega}_\nu^2 + z_3\bar{\omega}_\nu^3 + z_4\bar{\omega}_\nu^4\right). \tag{2.2}$$

Both formulae extend the definitions of (z_ν) and (ζ_ν) to periodic sequences of period 5. Since $\zeta_3 = \zeta_{-2}$, $\zeta_4 = \zeta_{-1}$, we may rewrite (2.1) as

$$z_\nu = \zeta_0 + \left(\zeta_1\omega_\nu + \zeta_{-1}\omega_\nu^{-1}\right) + \left(\zeta_2\omega_\nu^2 + \zeta_{-2}\omega_\nu^{-2}\right) \tag{2.3}$$

which is the so-called *real* f.F.s. of the (z_ν). Writing

$$\tilde{f}_\nu^1 = \zeta_1\omega_\nu + \zeta_{-1}\omega_\nu^{-1}, \qquad \tilde{f}_\nu^2 = \zeta_2\omega_\nu^2 + \zeta_{-2}\omega_\nu^{-2}, \tag{2.4}$$

we obtain the final form of the f.F.s. as

$$z_\nu = \zeta_0 + \tilde{f}_\nu^1 + \tilde{f}_\nu^2. \tag{2.5}$$

By (2.2) ζ_0 is the centroid of the z_ν. Selecting this centroid as the origin O of the complex plane, (2.5) simplifies to

$$z_\nu = \tilde{f}_\nu^1 + \tilde{f}_\nu^2 \qquad (\nu = 0,\dots,4). \tag{2.6}$$

Introducing the two new pentagons

$$\tilde{\Pi}^1 = \left(\tilde{f}_\nu^1\right) \quad \text{and} \quad \tilde{\Pi}^2 = \left(\tilde{f}_\nu^2\right), \tag{2.7}$$

we may represent the pentagon $\Pi = (z_\nu)$ in the form

$$\Pi = \tilde{\Pi}^1 + \tilde{\Pi}^2. \tag{2.8}$$

The simple nature of the pentagons (2.7) is shown by the following statements:

$\tilde{\Pi}^1$ *is an affine regular pentagon*, (2.9)

$\tilde{\Pi}^2$ *is an affine regular star-shaped pentagon*. (2.10)

A proof is immediate: Setting in the first relation (2.4) $\tilde{f}_\nu^1 = x_\nu + iy_\nu$, $\zeta_1 = a + bi$, $\zeta_{-1} = c + di$, we find that

$$x_\nu = (a+c)\cos\frac{2\pi\nu}{5} + (-b+d)\sin\frac{2\pi\nu}{5}$$

$$y_\nu = (b+d)\cos\frac{2\pi\nu}{5} + (a-c)\sin\frac{2\pi\nu}{5},$$

and (2.9) is established. Replacing in the right-hand sides ν by 2ν, we obtain (2.10).

So far we have made only general remarks on the f.F.s. of 5 terms which readily extend to the series for k terms. To obtain Theorem 1 we want to invert the real f.F.s. (2.6), i.e., find the individual terms $\tilde{f}^1$ and $\tilde{f}^2$. This is where Douglas's idea comes in. From (2.3), with $\zeta_0 = 0$, and writing $\omega = \omega_1$, we obtain

$$z_{\nu+2} = \left(\zeta_1\omega_\nu\omega^2 + \zeta_{-1}\omega_\nu^{-1}\omega^{-2}\right) + \left(\zeta_2\omega_\nu^2\omega^{-1} + \zeta_{-2}\omega_\nu^{-2}\omega\right),$$

$$z_{\nu-2} = \left(\zeta_1\omega_\nu\omega^{-2} + \zeta_{-1}\omega_\nu^{-1}\omega^2\right) + \left(\zeta_2\omega_\nu^2\omega + \zeta_{-2}\omega_\nu^{-2}\omega^{-1}\right),$$

and therefore

$$\begin{aligned} z'_\nu &= \tfrac{1}{2}(z_{\nu+2} + z_{\nu-2}) \\ &= \tfrac{1}{2}(\omega^2 + \omega^{-2})(\zeta_1\omega_\nu + \zeta_{-1}\omega_\nu^{-1}) + \tfrac{1}{2}(\omega + \omega^{-1})(\zeta_2\omega_\nu^2 + \zeta_{-2}\omega_\nu^{-2}). \end{aligned}$$

But then, by (2.4), we have

$$z'_\nu = \tilde{f}_\nu^1\cos\frac{4\pi}{5} + \tilde{f}_\nu^2\cos\frac{2\pi}{5}. \tag{2.11}$$

Since $\cos(4\pi/5) = -\cos(\pi/5)$, all that we have to do now is to invert the system of equations

$$\begin{aligned} z_\nu &= \tilde{f}_\nu^1 + \tilde{f}_\nu^2 \\ z'_\nu &= -\tilde{f}_\nu^1\cos\frac{\pi}{5} + \tilde{f}_\nu^2\cos\frac{2\pi}{5}. \end{aligned} \tag{2.12}$$

Since

$$\cos\left(\frac{\pi}{5}\right) = \frac{1+\sqrt{5}}{4}, \quad \cos\left(\frac{2\pi}{5}\right) = \frac{-1+\sqrt{5}}{4},$$

we readily find the solution of (2.12) to be given by

$$\tilde{f}_\nu^1 = \left(-\frac{1}{\sqrt{5}} z_\nu + \left(1 + \frac{1}{\sqrt{5}}\right) z'_\nu\right) \cdot \frac{1-\sqrt{5}}{2}$$

$$\tilde{f}_\nu^2 = \left(\frac{1}{\sqrt{5}} z_\nu + \left(1 - \frac{1}{\sqrt{5}}\right) z'_\nu\right) \cdot \frac{1+\sqrt{5}}{2}. \tag{2.13}$$

Introducing the new points

$$f_\nu^1 = -\frac{1}{\sqrt{5}} z_\nu + \left(1 + \frac{1}{\sqrt{5}}\right) z'_\nu,$$

$$f_\nu^2 = \frac{1}{\sqrt{5}} z_\nu + \left(1 - \frac{1}{\sqrt{5}}\right) z'_\nu, \tag{2.14}$$

we obtain the f.F.s. (2.6) in the form

$$z_\nu = \frac{1-\sqrt{5}}{2} f_\nu^1 + \frac{1+\sqrt{5}}{2} f_\nu^2. \tag{2.15}$$

Let us now establish Theorem 1 for the case where $\Pi \subset \mathbf{C}$. From the first relation (2.14) we find that

$$f_\nu^1 - z'_\nu = \frac{1}{\sqrt{5}} (z'_\nu - z_\nu), \tag{2.16}$$

while the second relation (2.14) shows that

$$f_\nu^2 - z'_\nu = -\frac{1}{\sqrt{5}} (z'_\nu - z_\nu). \tag{2.17}$$

(2.16), (2.17), are identical with the relations (1.3) that we wished to establish.

Why are the polygons Π^1 and Π^2, defined by (1.4) and (1.5), affine regular? From (2.13) and (2.14) we find that

$$f_\nu^1 = \tilde{f}_\nu^1 \Big/ \frac{1-\sqrt{5}}{2}, \quad f_\nu^2 = \tilde{f}_\nu^2 \Big/ \frac{1+\sqrt{5}}{2}, \tag{2.18}$$

while we know by (2.7), (2.9), (2.10) that the polygons $\tilde{\Pi}^1$ and $\tilde{\Pi}^2$ are affine regular. A proof of Theorem 1, for the case where $\Pi \subset \mathbb{C}$, follows from the relations (2.18).

3. A Proof of Theorem 1 if $\Pi \subset R^3$. We point out first that the definition of the pentagons (1.4) and (1.5), by the relations (1.2) and (1.3), remains valid in any real vector space, in particular for R^3. The only statements still in doubt are (2.9) and (2.10).

Let

$$F = (\Pi, \Pi^1, \Pi^2) \tag{3.1}$$

denote the space figure obtained by (1.2) and (1.3), and let

$$F_{xy} = \left(\Pi_{xy}, \Pi^1_{xy}, \Pi^2_{xy}\right) \qquad F_{xz} = \left(\Pi_{xz}, \Pi^1_{xz}, \Pi^2_{xz}\right) \tag{3.2}$$

be its orthogonal projections onto the coordinate planes xOy and xOz, respectively. Since the construction of F is affine invariant, it is clear that we can apply to the plane figures (3.2) the results of the last section, in particular

$$\textit{the pentagons } \Pi^1_{xy} \textit{ and } \Pi^1_{xz} \textit{ are affine regular.} \tag{3.3}$$

We now appeal to the following most elementary

LEMMA 1. *If the space pentagon*

$$\Pi^1 = (x_\nu, y_\nu, z_\nu) \qquad (\nu = 0,1,2,3,4) \tag{3.4}$$

has plane projections

$$\Pi^1_{xy} = (x_\nu, y_\nu), \quad \Pi^1_{xz} = (x_\nu, z_\nu) \tag{3.5}$$

which are affine regular pentagons, then Π^1 itself is a plane pentagon which is affine regular.

Proof: The affine regular pentagons (3.5) admit representations of the form

$$\begin{aligned} x_\nu &= a\cos\frac{2\pi\nu}{5} + b\sin\frac{2\pi\nu}{5}, \quad x_\nu = a'\cos\frac{2\pi\nu}{5} + b'\sin\frac{2\pi\nu}{5}, \\ y_\nu &= c\cos\frac{2\pi\nu}{5} + d\sin\frac{2\pi\nu}{5}, \quad z_\nu = e\cos\frac{2\pi\nu}{5} + f\sin\frac{2\pi\nu}{5}. \end{aligned} \tag{3.6}$$

On comparing the first two equations of (3.6) we conclude that we must have $a=a'$, $b=b'$, and so

$$x_\nu = a\cos\frac{2\pi\nu}{5} + b\sin\frac{2\pi\nu}{5},$$

$$y_\nu = c\cos\frac{2\pi\nu}{5} + d\sin\frac{2\pi\nu}{5},$$

$$z_\nu = e\cos\frac{2\pi\nu}{5} + f\sin\frac{2\pi\nu}{5}. \tag{3.7}$$

It follows that Π^1 is an affine regular pentagon in the plane defined by the oblique coordinate system of the two vectors $u=(a,c,e)$ and $v=(b,d,f)$. This completes our proof of Theorem 1.

REMARKS. 1. The two pentagons Π^1 and Π^2 of Theorem 1 lie in different planes but have as common center the centroid O of the vertices of Π. The problem of choosing Π so as to maximize the artistic effect of the entire structure is not mathematical and is, of course, hopeless.

2. Douglas's fortunate idea is to construct the pentagons Π^1 and Π^2 and not the pentagons

$$\tilde{\Pi}^1 = \frac{1-\sqrt{5}}{2}\Pi^1, \qquad \tilde{\Pi}^2 = \frac{1+\sqrt{5}}{2}\Pi^2 \tag{3.8}$$

which provide the final harmonic analysis

$$\Pi = \tilde{\Pi}^1 + \tilde{\Pi}^2 \tag{3.9}$$

according to (2.8). This idea simplifies considerably the final construction, because finding the pentagons (3.8) themselves would require two homothetic images with center O, a cumbersome complication.

4. The Graphical Harmonic Analysis of a Skew Heptagon. Our application of the f.F.s. to Douglas's theorem readily suggests the way to generalize his result to a closed skew polygon having k vertices. Having in mind further outdoor sculptures, we restrict our discussion to the case when $k=7$; hence

$$\Pi = (z_0, z_1, \ldots, z_6) \tag{4.1}$$

is a heptagon. We have omitted the case when $k=6$ for the reason that regular star-shaped hexagons are not particularly interesting. We commence our discussion by assuming that

$$\Pi \subset \mathbb{C} \tag{4.2}$$

when the z_ν are complex numbers. Their f.F.s. and its inverse formulae are

$$z_\nu = \sum_{\alpha=0}^{6} \zeta_\alpha \omega_\nu^\alpha, \quad \zeta_\nu = \frac{1}{7} \sum_{\alpha=0}^{6} z_\alpha \bar{\omega}_\nu^\alpha \qquad (\nu = 0,\ldots,6) \tag{4.3}$$

where $\omega_\nu = \exp(2\pi\nu/7)$. Again we assume that $z_0 + z_1 + \cdots + z_6 = 0$, hence $\zeta_0 = 0$, and folding the f.F.s., as in (2.3), we obtain

$$z_\nu = \left(\zeta_1\omega_\nu + \zeta_{-1}\omega_\nu^{-1}\right) + \left(\zeta_2\omega_\nu^2 + \zeta_{-2}\omega_\nu^{-2}\right) + \left(\zeta_3\omega_\nu^3 + \zeta_{-3}\omega_\nu^{-3}\right). \tag{4.4}$$

The midpoint of the side of Π that is opposite to the vertex z_ν is

$$z_\nu' = \tfrac{1}{2}\left(z_{\nu+3} + z_{\nu-3}\right). \tag{4.5}$$

However, now we also need the further midpoint

$$z_\nu'' = \tfrac{1}{2}\left(z_{\nu+2} + z_{\nu-2}\right). \tag{4.6}$$

From (4.4), and writing $\omega_1 = \omega$, we obtain

$$z_{\nu+3} = \left(\zeta_1\omega_\nu\omega^3 + \zeta_{-1}\omega_\nu^{-1}\omega^{-3}\right) + \left(\zeta_2\omega_\nu^2\omega^{-1} + \zeta_{-2}\omega_\nu^{-2}\omega\right)$$
$$+ \left(\zeta_3\omega_\nu^3\omega^2 + \zeta_{-3}\omega_\nu^{-3}\omega^{-2}\right)$$

and

$$z_{\nu-3} = \left(\zeta_1\omega_\nu\omega^{-3} + \zeta_{-1}\omega_\nu^{-1}\omega^3\right) + \left(\zeta_2\omega_\nu^2\omega + \zeta_{-2}\omega_\nu^{-2}\omega^{-1}\right)$$
$$+ \left(\zeta_3\omega_\nu^3\omega^{-2} + \zeta_{-3}\omega_\nu^{-3}\omega^2\right),$$

whence

$$z_\nu' = \frac{\omega^3 + \omega^{-3}}{2}\tilde{f}_\nu^1 + \frac{\omega + \omega^{-1}}{2}\tilde{f}_\nu^2 + \frac{\omega^2 + \omega^{-2}}{2}\tilde{f}_\nu^3, \tag{4.7}$$

if we write

$$\tilde{f}_\nu^j = \zeta_j\omega_\nu^j + \zeta_{-j}\omega_\nu^{-j} \qquad (j = 1,2,3;\ \nu = 0,\ldots,6). \tag{4.8}$$

Likewise we obtain from (4.4) that

$$z_{\nu+2} = \left(\zeta_1\omega_\nu\omega^2 + \zeta_{-1}\omega_\nu^{-1}\omega^{-2}\right) + \left(\zeta_2\omega_\nu^2\omega^{-3} + \zeta_{-2}\omega_\nu^{-2}\omega^3\right)$$
$$+ \left(\zeta_3\omega_\nu^3\omega^{-1} + \zeta_{-3}\omega_\nu^{-3}\omega\right)$$

and

$$z_{\nu-2} = \left(\zeta_1\omega_\nu\omega^{-2} + \zeta_{-1}\omega_\nu^{-1}\omega^2\right) + \left(\zeta_2\omega_\nu^2\omega^3 + \zeta_{-2}\omega_\nu^{-2}\omega^{-3}\right)$$
$$+ \left(\zeta_3\omega_\nu^3\omega + \zeta_{-3}\omega_\nu^{-3}\omega^{-1}\right),$$

whence

$$z_\nu'' = \frac{\omega^2+\omega^{-2}}{2}\tilde{f}_\nu^1 + \frac{\omega^3+\omega^{-3}}{2}\tilde{f}_\nu^2 + \frac{\omega+\omega^{-1}}{2}\tilde{f}_\nu^3. \tag{4.9}$$

By (4.4) and (4.8) we see that the real f.F.s. of Π is

$$z_\nu = \tilde{f}_\nu^1 + \tilde{f}_\nu^2 + \tilde{f}_\nu^3. \tag{4.10}$$

As in the case of pentagons, the analogue of Douglas's theorem will arise if we invert the 3×3 system of equations (4.10), (4.7), (4.9). Writing

$$\Omega_j = \frac{1}{2}\left(\omega^j + \omega^{-j}\right) = \cos\frac{2\pi j}{7} \qquad (j=1,2,3), \tag{4.11}$$

we are to solve the system

$$\begin{aligned} z_\nu &= \tilde{f}_\nu^1 + \tilde{f}_\nu^2 + \tilde{f}_\nu^3 \\ z_\nu' &= \Omega_3\tilde{f}_\nu^1 + \Omega_1\tilde{f}_\nu^2 + \Omega_2\tilde{f}_\nu^3 \\ z_\nu'' &= \Omega_2\tilde{f}_\nu^1 + \Omega_3\tilde{f}_\nu^2 + \Omega_1\tilde{f}_\nu^3. \end{aligned} \tag{4.12}$$

In terms of the inverse matrix

$$\begin{Vmatrix} A_1 & B_1 & C_1 \\ A_2 & B_2 & C_2 \\ A_3 & B_3 & C_3 \end{Vmatrix} = \begin{Vmatrix} 1 & 1 & 1 \\ \Omega_3 & \Omega_1 & \Omega_2 \\ \Omega_2 & \Omega_3 & \Omega_1 \end{Vmatrix}^{-1} \tag{4.13}$$

the solutions are

$$\tilde{f}_\nu^j = A_j z_\nu + B_j z_\nu' + C_j z_\nu'' \qquad (j=1,2,3). \tag{4.14}$$

By (4.8) it is clear that the three heptagons

$$\tilde{\Pi}^j = \left(\tilde{f}_0^j, \tilde{f}_1^j, \tilde{f}_2^j, \tilde{f}_3^j, \tilde{f}_4^j, \tilde{f}_5^j, \tilde{f}_6^j\right) \qquad (j=1,2,3), \tag{4.15}$$

are affine images of the three regular heptagons

$$\left(1,\omega,\omega^2,\omega^3,\omega^4,\omega^5,\omega^6\right), \quad \left(1,\omega^2,\omega^4,\omega^6,\omega,\omega^3,\omega^5\right), \tag{4.16}$$
$$\left(1,\omega^3,\omega^6,\omega^2,\omega^5,\omega,\omega^4\right),$$

respectively. In terms of the heptagons (4.15) we may write (4.10) as

$$\Pi = \tilde{\Pi}^1 + \tilde{\Pi}^2 + \tilde{\Pi}^3. \tag{4.17}$$

However, the heptagons (4.15) are *not* the ones that we wish to construct. Rather, following Douglas's lead, we introduce the weights

$$\alpha_j = \frac{A_j}{s_j}, \quad \beta_j = \frac{B_j}{s_j}, \quad \gamma_j = \frac{C_j}{s_j}, \quad \text{where} \quad s_j = A_j + B_j + C_j, \tag{4.18}$$

and want to construct the heptagons,

$$\Pi^j = (f_0^j, f_1^j, f_2^j, f_3^j, f_4^j, f_5^j, f_6^j) \qquad (j=1,2,3), \tag{4.19}$$

having vertices given by

$$f_\nu^j = \alpha_j z_\nu + \beta_j z'_\nu + \gamma_j z''_\nu \qquad (j=1,2,3). \tag{4.20}$$

We state our results as

THEOREM 2. *Let*

$$\Pi = (z_0, z_1, \ldots, z_6) \tag{4.21}$$

be a skew heptagon in R^3, *and let*

$$z'_\nu = \tfrac{1}{2}(z_{\nu+3} + z_{\nu-3}), \quad z''_\nu = \tfrac{1}{2}(z_{\nu+2} + z_{\nu-2}) \tag{4.22}$$

be the midpoints of appropriate sides and chords of Π. *By* (4.11), (4.13), *and* (4.18) *we define the three sets of numerical weights*

$$\alpha_j, \beta_j, \gamma_j, \quad \alpha_j + \beta_j + \gamma_j = 1 \qquad (j=1,2,3). \tag{4.23}$$

In each of the seven triangles

$$T_\nu = (z_\nu, z'_\nu, z''_\nu) \qquad (\nu=0,\ldots,6), \tag{4.24}$$

we define the three points

$$f_\nu^1, f_\nu^2, f_\nu^3 \tag{4.25}$$

as the centroids of T_ν *with the three sets of weights* (4.23), *respectively. Equivalently,* (4.25) *are defined by the equations* (4.20). *Then the three heptagons*

$$\Pi^j = (f_0^j, f_1^j, f_2^j, f_3^j, f_4^j, f_5^j, f_6^j) \qquad (j=1,2,3) \tag{4.26}$$

are plane heptagons and they are affine images of the regular heptagons (4.16), *respectively.*

Our Theorem 2 is, of course, fully established if we assume that $\Pi \subset \mathbb{C}$. That it remains true if $\Pi \subset R^3$ follows from reasonings similar to those used in extending Theorem 1 from R^2 to R^3, in particular from the lemma: *If a heptagon* Π *in* R^3 *has two affine regular plane projections, then* Π *itself is plane and affine regular.*

5. The Construction of a Space Model Illustrating Theorem 2. By this we mean the construction of the figure

$$F = (\Pi, \Pi^1, \Pi^2, \Pi^3), \tag{5.1}$$

where Π, Π^1, Π^2, Π^3 are the heptagons of Theorem 2. This could be done graphically on a sheet of paper by the methods of Descriptive Geometry. However, we have in mind a 3-dimensional structure made out of thin (wooden) sticks.

For this purpose we need the numerical values of the weights (4.18). With sufficient accuracy for any physical construction, these are as follows:

$$\left\| \begin{matrix} \alpha_1 & \beta_1 & \gamma_1 \\ \alpha_2 & \beta_2 & \gamma_2 \\ \alpha_3 & \beta_3 & \gamma_3 \end{matrix} \right\| = \left\| \begin{matrix} -.08627 & .69859 & .38768 \\ .78485 & 1.08626 & -.87111 \\ .30141 & .21515 & .48344 \end{matrix} \right\| \tag{5.2}$$

$$s_1 = -1.24697, \quad s_2 = .44504, \quad s_3 = 1.80193. \tag{5.3}$$

The construction of the 14 points z_ν' and z_ν'' by the formulae (4.22) presents no difficulties. These also determine the 7 triangles (4.24).

In the plane of each T_ν we are now to construct the centroids (4.25) for the three sets of weights (4.23). Here we use the following lemma, which is too elementary to require a proof. (The reader is asked to supply a diagram.)

LEMMA 2. *Let*

$$T = (z, z', z'') \tag{5.4}$$

be a triangle, and let

$$f = \alpha z + \beta z' + \gamma z'' \tag{5.5}$$

be its centroid for the weights α, β, γ, *with* $\alpha + \beta + \gamma = 1$.

If h *denotes the intersection of the line joining* z *to* z', *with the line joining* z'' *to* f, *then the relations*

$$h - z' = \rho(z' - z), \quad f - h = \sigma(h - z'') \tag{5.6}$$

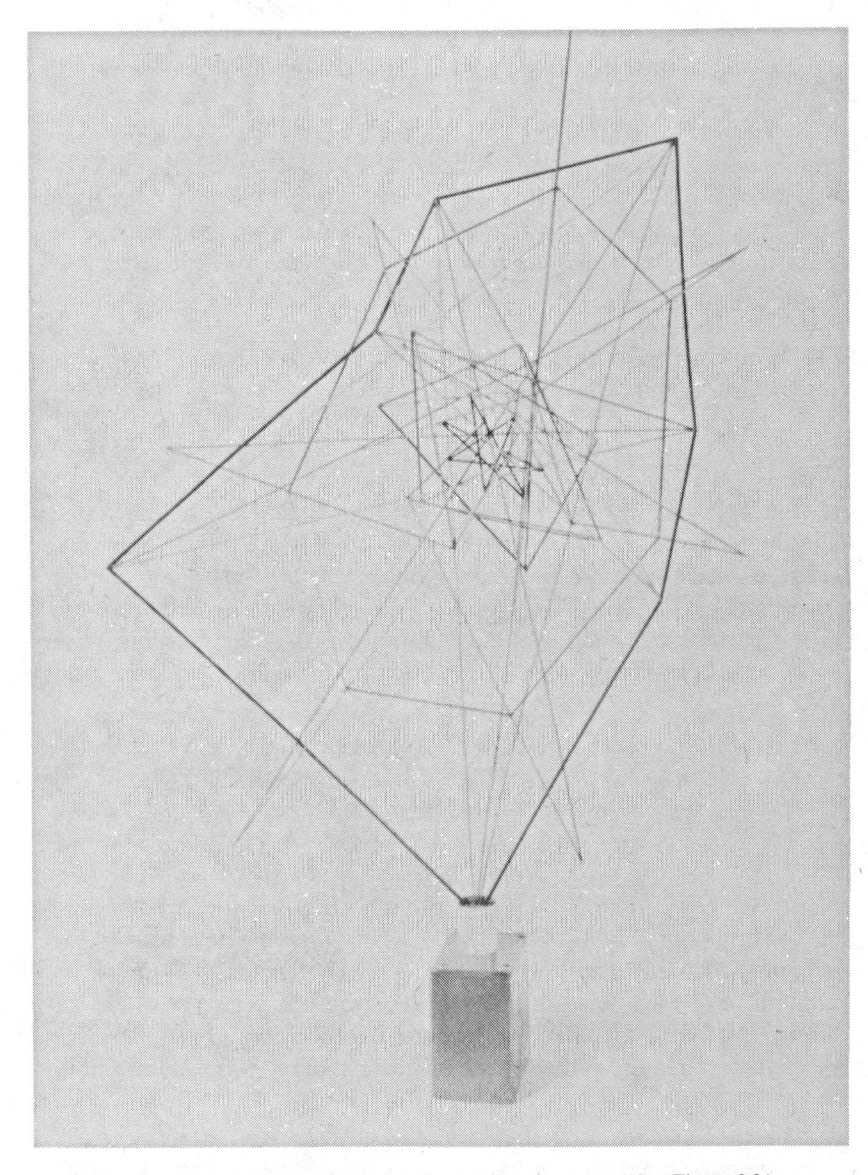

A model of the harmonic analysis of a skew heptagon. (See Figure 9.3.)

hold, where

$$\rho = -\frac{\alpha}{\alpha+\beta}, \quad \sigma = -\gamma. \tag{5.7}$$

We apply Lemma 2 to each T_ν with the sets of weights (5.2). We drop the subscript ν and show in Figure 9.2 the location of the centroids f^1, f^2, f^3 in the plane of the triangle $T=(z, z', z'')$. Using Lemma 2 and the numerical values (5.2), we obtain the relations

$$\begin{aligned} h^1 - z' &= \rho_1(z'-z), \quad f^1 - h^1 = \sigma_1(h^1 - z'') \\ h^2 - z' &= \rho_2(z'-z), \quad f^2 - h^2 = \sigma_2(h^2 - z'') \\ h^3 - z' &= \rho_3(z'-z), \quad f^3 - h^3 = \sigma_3(h^3 - z''). \end{aligned} \tag{5.8}$$

The numerical values of the ratios ρ_j and σ_j, given by (5.7) and (5.2), are

$$\begin{aligned} \rho_1 &= .14089, & \sigma_1 &= -.38768 \\ \rho_2 &= -.41946, & \sigma_2 &= .87111 \\ \rho_3 &= -.58350, & \sigma_3 &= -.48344. \end{aligned} \tag{5.9}$$

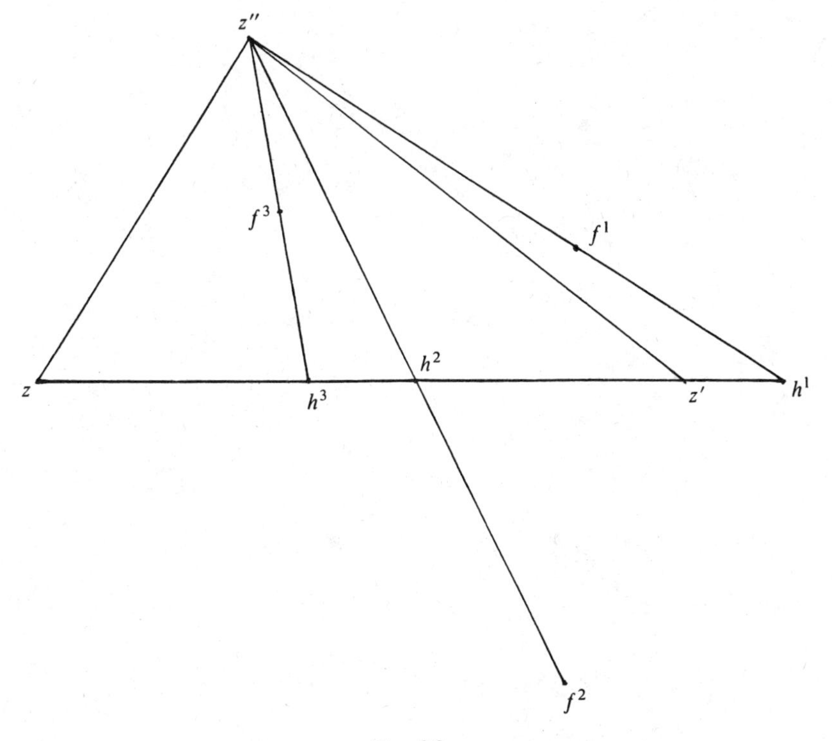

FIG. 9.2

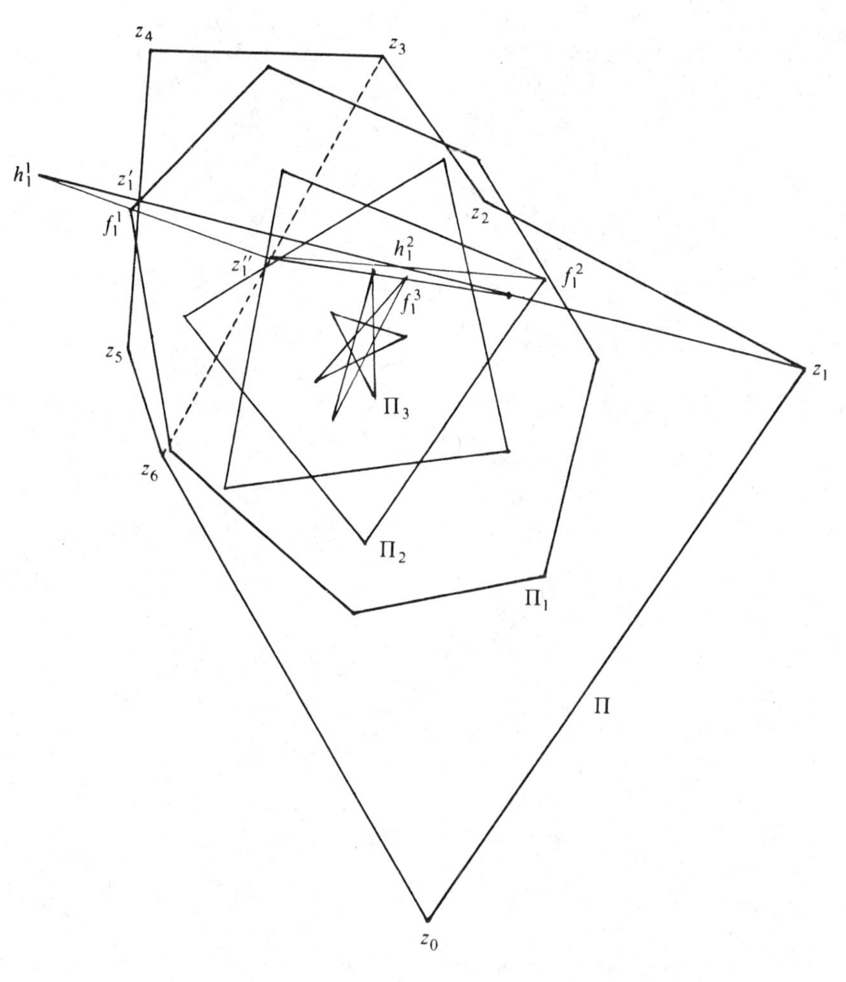

FIG. 9.3

The locations of the points h^j and f^j in Figure 9.2 are drawn to scale. For any other triangle $T_\nu = (z_\nu, z'_\nu, z''_\nu)$ the corresponding diagram is the image of Figure 9.2 by the affine transformation mapping T onto T_ν.

Our Figure 9.3 shows a 2-dimensional illustration of Theorem 2. It shows the three affine regular heptagons Π^1, Π^2, and Π^3. In order to simplify the drawing it shows only the construction of the three vertices

$$f_1^1, f_1^2, f_1^3,$$

corresponding to the triangle $T_1 = (z_1, z'_1, z''_1)$.

Concluding remark. The reader has noticed the important role played by affine maps. The harmonic components of closed skew polygons are *not* regular polygons, but rather the *affine maps* of such. It seems too tempting not to mention here that the old masters of painting were familiar with affine maps and that they used them for the peculiar purpose of concealing a human skull, especially in portraits, to remind the sitter of the transitory nature of human life. A classic example is the portrait of two French ambassadors by Holbein the Younger. "The Ambassadors" is in the National Gallery in London and shows in the lower part of the painting an object in the shape of a long, slanting log. It actually represents a tremendously elongated human skull and can be recognized as such if the viewer leaves the room by the door next to the painting and takes a last look at it which will foreshorten the skull to recognizable proportions. The technical term for this device is "anamorphism."

Problems

The first two problems below are from the prepint [3], where also references are given.

1. Let $\Pi = (z_0, z_1, z_2, z_3)$ be a plane parallelogram having its vertices z_ν in cyclic order. On its sides we erect externally four squares and let s_0, s_1, s_2, s_3 be their centers. Show that $s_0 s_1 s_2 s_3$ is a square.

2. Let $S = (z_0, z_1, z_2, z_3)$ and $S' = (z'_0, z'_1, z'_2, z'_3)$ be two squares of opposite orientation in the plane having a common vertex, $z_0 = z'_0$, say. Show that the centers of S and S' and the midpoints of the segments $[z_1, z'_1]$, $[z_3, z'_3]$ form the vertices of a square.

The next problem is taken from [**1**, first example on p. 125].

3. Let $\Pi = (z_\nu, \nu = 0, \ldots, 4)$ be a skew pentagon in R^3. With the vertex z_ν we associate the midpoint

$$z_\nu^* = \tfrac{1}{2}(z_{\nu+1} + z_{\nu-1})$$

of its two neighboring vertices. Determine the points g^1 and g^2 from the relations

$$g_\nu^1 - z_\nu^* = -\frac{1}{\sqrt{5}}(z_\nu^* - z_\nu), \quad g_\nu^2 - z_\nu^* = \frac{1}{\sqrt{5}}(z_\nu^* - z_\nu) \qquad (\nu=0,\dots,4).$$

Show that $\Gamma^1 = (g_\nu^1)$ is an affine regular pentagon and that $\Gamma^2 = (g_\nu^2)$ is an affine regular star-shaped pentagon.

(*Hint*: Assume first that Π is in the complex plane. Use the f.F.s. (2.3), with $\zeta_0 = 0$, to show that in the notation of the relations (2.14), we have the relations

$$g_\nu^1 = -f_\nu^1, \quad g_\nu^2 = -\tfrac{1}{2}\left(3+\sqrt{5}\right)f_\nu^2 \qquad (\nu=0,\dots,4).$$

These show that Γ^1 and Γ^2 are homothetic images of Π^1 and Π^2, respectively, and our problem is solved.)

4. Let Π be a closed skew pentagon in R^3. Show that Douglas's second star-shaped pentagon Π^2 is contained within the convex hull $K(\Pi)$ of the pentagon Π.

5. Let Π be a closed skew heptagon in R^3. Show that the third star-shaped heptagon Π^3, of Theorem 2, is contained within the convex hull $K(\Pi)$.

References

[1] Jesse Douglas, *Geometry of polygons in the complex plane*, J. Math. Phys., 19 (1940) 93–130.

[2] ______, *A theorem on skew pentagons*, Scripta Math., 25 (1960) 5–9.

[3] J. C. Fisher, D. Ruoff, and J. Shilleto, *Polygons and polynomials*, Preprint No. 32, May 1979, University of Regina, Regina, Canada.

[4] I. J. Schoenberg, *The finite Fourier series and elementary geometry*, Amer. Math. Monthly, 57 (1950) 390–404.

[5] ______, *The harmonic analysis of skew polygons as a source of outdoor sculptures*, submitted to the Proceedings of the Coxeter Symposium, May 21 to May 25, 1979, University of Toronto. To appear in *The Geometric Vein*, Springer, New York, 1982.

CHAPTER **10**

SPLINE FUNCTIONS AND ELEMENTARY NUMERICAL ANALYSIS

1. Introduction. By elementary numerical analysis we mean the methods, pioneered by Newton, of obtaining numerical approximations for the interpolation, integration, differentiation, and other such functionals, of functions of one variable. In these methods interpolation plays a preferred and fundamental role. Starting in 1949 Arthur Sard introduced a new point of view leading to his *best* approximation formulae. In this new approach to Newton's old problems it seemed as if interpolation was losing the fundamental role that it played in Newton's work. The purpose of this chapter is to show that we again restore interpolation to its central role, provided that we replace the Newtonian polynomial interpolation of the data by interpolation by the so-called natural spline functions.

This chapter is a slightly amplified version of the first of four lectures given at Swarthmore College in 1966, which were well recorded and edited by Steven Maurer, of the class of 1967 [**5**]. It is a survey lecture and no proofs are given, but we give precise definitions and state the main theorems. For their proofs we refer to Greville's excellent paper [**2**]. No previous knowledge of numerical analysis, elementary or otherwise, is required.

2. A New Approach to Elementary Numerical Analysis. Today I am speaking about a new approach to elementary numerical analysis (ENA) with the emphasis on elementary. Perhaps I may start with a little story. The emperor William I of Germany dropped in at the Potsdam Observatory sometime in the 1880's and spoke to a famous resident astronomer whose name escapes me at the moment. The emperor asked him, "Professor, what's new in heaven?" The professor answered, "Majesty, I presume you know everything old about the heavens." Returning to our subject, it would be very nice if all of you in the audience would know all about ENA—the old one. But fortunately, ENA is not as vast as the heavens, and I might undertake to give in this hour most of what is old in heaven and also something new. Our mention of the heavens does not seem inappropriate, since Newton's first important application of interpolation was to the location of comets from a few observations.

The new approach has much to do with a new type of very elementary functions which are called spline functions. I may as well start by describing them. As you know, the simplest functions in pure and applied mathematics are polynomials. Let π_k denote the class of polynomials of degree not exceeding k. First there is π_0, the constants, not very interesting. Then comes π_1, the straight lines, and next we have π_2, the parabolas. With graphs and problems, this is already most of high school mathematics. Then we have the more exciting cubics π_3, and so on. When we mention polynomials we must mention the theorem of Weierstrass which says that polynomials will imitate within any prescribed tolerance any kind of continuous curve you may wish to prescribe in a compact range. This is the theoretical foundation of the use of polynomials.

Beyond polynomials we have the so-called step functions, which are clearly related to π_0. If we take the integral of a step function we get a broken linear function. If we integrate again, we get a succession of parabolic arcs. Notice that the order of continuity increases with each integration. We started with a function with discontinuities; with one integration we got a continuous function, and after two integrations we have a function with a continuous derivative. These are already spline functions. Generally, by a spline function S_m we mean a succession of polynomial arcs of degree not exceeding m, which are joined at certain points called knots, at which the order of continuity is as high as it can be without the two arcs being part of the same polynomial. In other words, a spline function S_m differs from a polynomial P_m of degree m in that both have m derivatives, but while the mth derivative of a polynomial is a constant, the mth derivative of a spline function is a step function.

Before we leave this perfunctory description of spline functions—in fact, this is all there is to it—let us briefly discuss the analytic representation of splines (we use also the term "spline" as a noun). For this it is useful to introduce the "plus function" x_+ defined by

$$x_+ = \max(x,0) = \tfrac{1}{2}(|x|+x) = \begin{cases} x & \text{if } x \geqslant 0, \\ 0 & \text{if } x < 0. \end{cases}$$

A broken linear function can now be described nicely using the plus function. If $x_1 < x_2 < \cdots < x_n$ are the points at which the broken linear function S_1 changes its slope, then with appropriate constants

$$S_1(x) = a_0 x + a_1 + c_1(x-x_1)_+ + c_2(x-x_2)_+ + \cdots + c_n(x-x_n)_+ .$$

If we integrate $S_1(x)$ we get a quadratic spline $S_2(x)$. Since

$$\int x_+ \, dx = \tfrac{1}{2}x_+^2 + c$$

we see that an arbitrary quadratic spline $S_2(x)$ is

$$S_2(x) = \tfrac{1}{2}a_0x^2 + a_1x + a_2 + \sum_{k=1}^{n} c_k(x-x_k)_+^2/2.$$

In general

$$S_m(x) = P_m(x) + \sum_{k=1}^{n} c_k(x-x_k)_+^m/m!,$$

where $P_m(x)$ is an arbitrary polynomial of degree m. We see that derivatives of splines are again splines and also that integrals of splines are splines. This is why splines are so useful in Calculus.

Now let us turn to ENA and see what some of the fundamental problems are. The main problem is to approximate a linear functional, of a function f, from given data about f. That is, we have a certain linear functional $\mathfrak{L}f$ and we want to approximate it from the data, which are values of f at certain points. It is clear that we can only *approximate* $\mathfrak{L}f$ if the values of f are measurements obtained by experiments. However, we are often forced to approximate even for mathematical functions. For example, suppose we wish to compute

$$\int_0^1 e^{-x^2}dx.$$

This is a problem in ENA rather than calculus because, as we all know, e^{-x^2} cannot be integrated in closed form. So what we do is compute e^{-x^2} at various points and then apply the methods of ENA.

What are these methods of ENA? Or, referring to our little story, what is old in heaven? I would say that these problems were solved for the first time —and for the last time—by Newton, because as far as I know that is all there is to the classical approach to ENA. I will limit myself to the case where we are given the values $f(x_i)$ $(i=1,2,\ldots,n)$, where $a \leqslant x_1 < x_2 < \cdots < x_n \leqslant b$. These I call data of the Lagrange type. If some derivatives are also given, they would be of the Hermite type. Now, Newton proceeds as follows: He interpolates the $f(x_i)$ by a polynomial $P(x) \in \pi_{n-1}$. This can always be done and

$$P(x) = \sum_{i=1}^{n} \frac{\omega(x)}{(x-x_i)\omega'(x_i)} f(x_i),$$

where

$$\omega(x) = \prod_{i=1}^{n} (x-x_i).$$

This is the famous interpolation formula of Lagrange. Note that the

$$\frac{\omega(x)}{(x-x_i)\omega'(x_i)} = L_i(x) \qquad (i=1,\dots,n)$$

are just the polynomials which assume the value 1 at x_i and 0 at the $n-1$ other points: hence

$$L_i(x_j) = \begin{cases} 1 & \text{if } i=j, \\ 0 & \text{if } i\neq j. \end{cases}$$

The $L_i(x)$ are called the *fundamental functions* of the interpolation problem, because we may write

$$P(x) = \sum_{i=1}^{n} f(x_i)L_i(x).$$

Let $f \approx g$ mean that g is a good approximation of f. Then, according to Newton,

$$f(x) \approx P(x) = \sum_{1}^{n} f(x_i)L_i(x),$$

that is, $P(x)$ is a good approximation of $f(x)$. If we want to differentiate, Newton writes

$$f'(x) \approx P'(x) = \sum_{1}^{n} f(x_i)L_i'(x).$$

Or, if we want to integrate, then

$$\int_a^b f(x)\,dx \approx \int_a^b P(x)\,dx = \sum_{1}^{n} f(x_i)A_i \quad \text{where} \quad A_i = \int_a^b L_i(x)\,dx.$$

Generally, for any kind of linear functional $\mathfrak{L}f$, according to Newton

$$\mathfrak{L}f \approx \mathfrak{L}P = \sum_{1}^{n} f(x_i)B_i \quad \text{where} \quad B_i = \mathfrak{L}L_i. \tag{2.1}$$

Notice *the fundamental role of interpolation*. Once you have interpolated you can do anything else because you perform it on the interpolating polynomial. Notice that all these formulae have a common property:

> *If $f \in \pi_{n-1}$, then they all give exact values; that is, $\approx$ can be replaced by $=$ in all cases, because $f(x) = P(x)$ for all x.*

This is already a big chunk of practical numerical analysis. But let us go on. *How can we get the* B_i *of* (2.1) *directly, without using the interpolating polynomial*? The B_i can be determined directly from (2.1) if we make sure that (2.1) is an equality for all $f \in \pi_{n-1}$. This will be the case, provided that the B_i are the solutions of the set of linear equations

$$\mathfrak{L}x^r = \sum_{i=1}^{n} B_i x_i^r \qquad (r=0,1,\ldots,n-1). \tag{2.2}$$

This is a system of n equations in n unknowns with the nonvanishing determinant known as the Vandermonde determinant. With the B_i satisfying (2.2) we can always write

$$\mathfrak{L}f = \sum_{1}^{n} B_i f(x_i) + Rf, \tag{2.3}$$

where Rf is the *remainder*, which of course will not vanish if $f \notin \pi_{n-1}$.

If $f \notin \pi_{n-1}$, it is important to get an idea about the size of the error in using the approximation (2.1); that is, we need an estimate of the remainder term Rf in (2.3). In this respect there is a very important contribution, the theorem of G. Peano of 1913, of which I wish to speak. I refer to formula (2.3), where I assume that the B_i are already determined so as to satisfy (2.2). We can solve (2.3) for the remainder and write

$$Rf = \mathfrak{L}f - \sum_{1}^{n} B_i f(x_i). \tag{2.4}$$

This remainder is also a linear functional, and it has the important property that

$$Rf = 0 \quad \text{if} \quad f \in \pi_{n-1}. \tag{2.5}$$

Suppose that there is a certain function $K(x)$ so that for all f we have

$$Rf = \int_a^b K(x) f^{(n)}(x)\, dx. \tag{2.6}$$

If Rf happens to be of the form (2.6), then it certainly enjoys the property (2.5); for if $f \in \pi_{n-1}$, then $f^{(n)}(x)$ vanishes identically. Therefore (2.6) *implies* (2.5). Peano's theorem is the converse: (2.5) *implies* (2.6). It states that if (2.5) holds, then we can express Rf in the form (2.6) with a kernel $K(x)$ which can be written explicitly as

$$K(x) = R_t \frac{(t-x)_+^{n-1}}{(n-1)!}. \tag{2.7}$$

Here we have two variables, and the subscript t of R means that the remainder operation, defined by (2.4), is to be performed with respect to the variable t. Now these are important, elegant, and elementary developments. You find them fully discussed, with many examples and applications, in Philip Davis's book [**1**].

The representation (2.6) is convenient for estimating the remainder. From the Schwarz-Buniakowski inequality applied to (2.6) we obtain that

$$(Rf)^2 \leqslant \left(\int_a^b (K(x))^2\,dx\right)\left(\int_a^b (f^{(n)}(x))^2\,dx\right). \tag{2.8}$$

If we have some information on $f^{(n)}(x)$, then (2.8) will provide useful estimates of Rf. This completes our description of the Newtonian approach to ENA.

Now, what is the new approach to ENA? The new approach was begun by Arthur Sard in about 1949. Sard goes beyond Newton in the following way. Above we determined the B_i by the n equations (2.2), and it would seem that we need n equations because we have n parameters B_i to be determined. Sard chooses an integer m such that

$$1 \leqslant m < n \tag{2.9}$$

and requires that not the full set of equations (2.2) be satisfied, but only

$$\mathfrak{L}x^r = \sum_{i=1}^{n} B_i x_i^r \qquad (r = 0, 1, \ldots, m-1). \tag{2.10}$$

From this set of m equations we can express $B_1, B_2, \ldots, B_m$, in terms of the $n-m$ parameters

$$B_{m+1}, B_{m+2}, \ldots, B_n \tag{2.11}$$

which as yet remain free.

Sard's best approximation of $\mathfrak{L}f$. With the coefficients B_i satisfying (2.10) we define again the remainder functional

$$Rf = \mathfrak{L}f - \sum_1^n B_i f(x_i) \tag{2.12}$$

which now depends on the $n-m$ arbitrary parameters (2.11). This entire family of remainders has, in view of (2.10), the property that

$$Rf = 0 \quad \text{if} \quad f \in \pi_{m-1}. \tag{2.13}$$

But then we can apply to Rf the Peano theorem and express it in the form

$$Rf = \int_a^b K(x) f^{(m)}(x)\, dx, \tag{2.14}$$

where the kernel $K(x)$ still depends on the arbitrary parameters (2.11). Sard determines these parameters by the condition that

$$\int_a^b (K(x))^2\, dx = \text{minimum}. \tag{2.15}$$

This requirement determines all coefficients B_i, and with their values so obtained, Sard defines

$$\sum_1^n B_i f(x_i) \tag{2.16}$$

as *the best approximation of order m of the functional $\mathfrak{L}f$*. Rewriting (2.12) in the form

$$\mathfrak{L}f = \sum_1^n B_i f(x_i) + Rf \tag{2.17}$$

we again have for its remainder, this time by (2.14), the inequality

$$(Rf)^2 \leqslant \int_a^b (K(x))^2\, dx \cdot \int_a^b (f^{(m)}(x))^2\, dx. \tag{2.18}$$

This is the origin of Sard's idea and explains why he minimizes the first factor on the right-hand side.

It should be clear that minimizing the quadratic function (2.15), under the side conditions (2.10), is numerically a rather laborious affair. However, it can be done, and Sard wrote his book [**3**] devoted to these problems. Sard's approach seems complicated if compared to Newton's approach, in which the special role of interpolation largely simplified matters.

The following question suggests itself: *Is there some interpolation procedure underlying Sard's method of approximating functionals*? Is there an interpolation method which would allow us to find approximations of functionals, in the same way as Newton did, by simply operating with $\mathfrak{L}$ on the interpolating function? The answer is yes, there is one, and it is the subject of our next section.

3. Natural Spline Interpolation. We begin with some definitions. We are given n points

$$x_1 < x_2 < \cdots < x_n \qquad (n>1), \tag{3.1}$$

and select an integer m such that

$$1 \leqslant m < n. \tag{3.2}$$

By a *natural spline of degree* $2m-1$ *having the knots* (3.1), we mean a function $S(x)$ with the following three properties:

$$S(x) \in \pi_{2m-1} \text{ in each interval } (x_i, x_{i+1}) \qquad (i=1,\dots,n-1), \tag{3.3}$$

$$S(x) \in C^{2m-2}, \tag{3.4}$$

$$S(x) \in \pi_{m-1} \text{ in } (-\infty, x_1) \text{ and also in } (x_n, +\infty). \tag{3.5}$$

We denote by

$$\mathcal{S}_{2m-1}(x_1,\dots,x_n) \tag{3.6}$$

the class of these splines. Note that $\pi_{m-1} \subset \mathcal{S}_{2m-1}(x_i)$.

Let us look at the simplest examples.

Let $m=1$. The conditions (3.3), (3.4), (3.5) show that the graph of $S(x)$ is a continuous polygonal line having vertices at the knots (3.1) and is such that $S(x)=S(x_1)$ if $x<x_1$, and $S(x)=S(x_n)$ if $x>x_n$ (Figure 10.1).

Let $m=2$. Again our conditions show that $S(x)$ is a cubic spline, therefore having continuous first and second derivatives, with knots (3.1), and the peculiar property that the two infinite branches of the graph are two straight lines (Figure 10.2).

If $m=3$, then $S(x)$ is a quintic spline, hence of class $C^{(4)}$, with its infinite branches being quadratics, a.s.f.

There is an analytic representation of $S(x)$. Let us look at the case $m=2$ and assume that

$$S(x) \in \mathcal{S}_3(x_1,\dots,x_n). \tag{3.7}$$

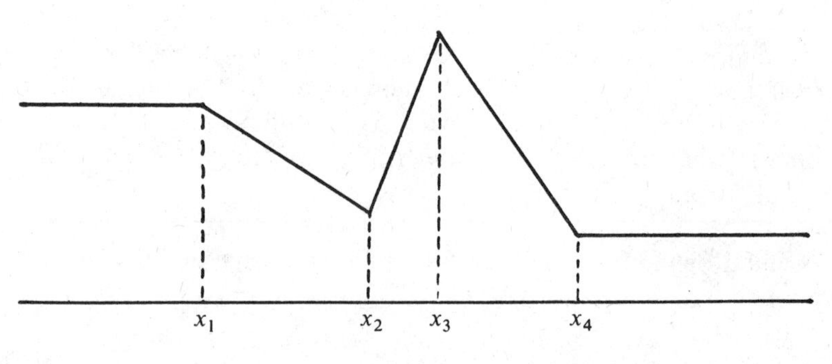

FIG. 10.1

From the first condition (3.5) we obtain the representation

$$S(x) = a_0 x + a_1 + c_1(x-x_1)_+^3 + \cdots + c_n(x-x_n)_+^3 \tag{3.8}$$

where the second condition (3.5) has as yet not been met. To enforce it, let $x > x_n$, and observe that we may therefore drop in (3.8) the subscripts "+". Expanding the binomials and collecting terms, we see that $S(x)$ will be linear in (x_n, ∞), provided that we have

$$c_1 + c_2 + \cdots + c_n = 0, \qquad c_1 x_1 + c_2 x_2 + \cdots + c_n x_n = 0. \tag{3.9}$$

Counting parameters we see that (3.8) depends on $2+n$ parameters, two of which can be eliminated by the two equations (3.9). *We are therefore left with n linear parameters.* Also in the general case when

$$S(x) \in \mathbb{S}_{2m-1}(x_i) \tag{3.10}$$

we find that

$$S(x) \textit{ depends on } m + n - m = n \textit{ linear parameters.} \tag{3.11}$$

At this point we suspect that $S(x)$ can interpolate uniquely n given ordinates at the knots. This is

THEOREM 1 (Natural spline interpolation). *Let the inequalities* (3.2) *hold. Given the reals*

$$y_1, y_2, \ldots, y_n, \tag{3.12}$$

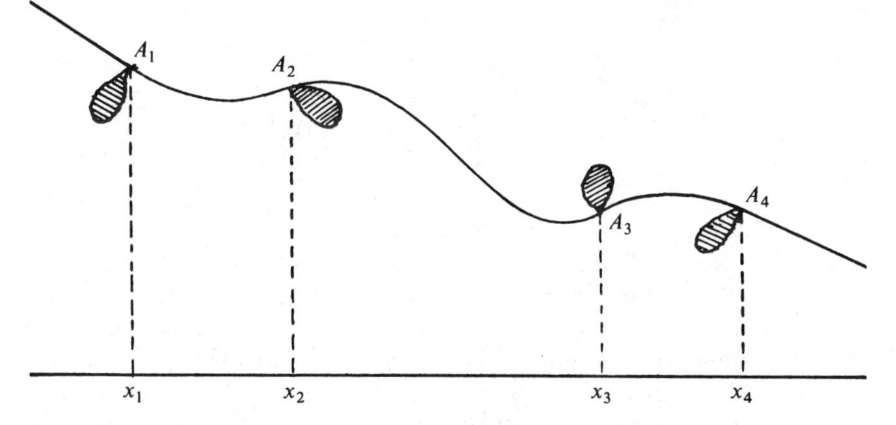

FIG. 10.2

there is a unique

$$S(x) \in \mathbb{S}_{2m-1}(x_1,\ldots,x_n) \tag{3.13}$$

such that

$$S(x_i) = y_i \qquad (i=1,\ldots,n). \tag{3.14}$$

Our interpolation method being linear, we may consider its fundamental functions $L_i(x)$ defined by

$$L_i(x) \in \mathbb{S}_{2m-1}(x_i), \qquad L_i(x_j) = \begin{cases} 1 & \text{if } i=j, \\ 0 & \text{if } i \neq j. \end{cases} \tag{3.15}$$

They allow us to write the interpolant (3.13) as

$$S(x) = \sum_1^n f(x_i) L_i(x). \tag{3.16}$$

If

$$f(x) \in \mathbb{S}_{2m-1}(x_i) \tag{3.17}$$

then the unicity in Theorem 1 implies that $S(x) = f(x)$ for all x. However, if (3.17) does not hold, then we may write

$$f(x) = \sum_1^n f(x_i) L_i(x) + R_f(x) \tag{3.18}$$

with a remainder $R_f(x)$. We call (3.18) the *natural spline interpolation formula* (abbreviated to N.S.I.F.).

COROLLARY 1. *In* (3.18) *we have*

$$R_f(x) = 0 \quad \textit{if} \quad f \in \mathbb{S}_{2m-1}(x_1,\ldots,x_n). \tag{3.19}$$

The name spline interpolation seems justified because we interpolate by splines. But *where does the adjective* "natural" *come from*? This will be explained and also the origin of the name "spline," if we return for a moment to the case $m=2$ of Figure 10.2. A spline is an old tool used by engineers to draw smooth curves that are to pass through given points A_1, $A_2, \ldots$. A spline is a thin, straight, flexible rod that is forced to pass through these points by being held at these points by special heavy objects called "dogs." If the points A_i are nearly on a horizontal line, then the linearized theory of the elastic beam, due to Euler and D. Bernoulli, shows that the resulting curve is the graph of a cubic polynomial between any two consecutive A_i and A_{i+1}. Moreover, these cubic arcs join with two continuous derivatives. Finally, the two end-pieces of the spline are *straight*, because there are no forces acting on them. This is the origin of the condition (3.5) for $m=2$, and of the term "natural." It does seem remarkable that this simple physical situation should generalize to Theorem 1.

4. Natural Spline Interpolation and Sard's Best Approximation Formulae. The N.S.I.F. (3.18) will restore the interpolation problem to the role that it played in Newton's case. Let $\mathfrak{L}f$ be a prescribed functional of the form

$$\mathfrak{L}f = \int_a^b \{a_0(x)f(x) + a_1(x)f'(x) + \cdots + a_{m-1}(x)f^{(m-1)}(x)\}\,dx$$

$$+ \sum_{i=1}^{j_0} b_{i0} f(x_{i0}) + \cdots + \sum_{i=1}^{j_{m-1}} b_{i,m-1} f^{(m-1)}(x_{i,m-1}). \tag{4.1}$$

The simplest examples of such functionals are

$$\int_a^b f(x)\,dx, \quad f'(x), \quad f(x), \tag{4.2}$$

where the last means that we are to interpolate the data $f(x_i)$ at x.

This much is clear: As the relation

$$f(x) = \sum_1^n f(x_i)L_i(x) \tag{4.3}$$

is an identity in x whenever $f \in \mathcal{S}_{2m-1}(x_i)$, we can operate with $\mathfrak{L}$ on both sides of (4.3) and get the equation

$$\mathfrak{L}f = \sum_1^n f(x_i)\mathfrak{L}L_i(x), \tag{4.4}$$

which is again valid if $f \in \mathcal{S}_{2m-1}(x_i)$.

However, it does seem remarkable that in this way we obtain the best approximation of order m of $\mathfrak{L}f$ in the sense of Sard. This is our

THEOREM 2. *In order to obtain the best approximation* (2.16) *of* $\mathfrak{L}f$ *in the sense of Sard, we need only operate with* $\mathfrak{L}$ *on both sides of the natural spline interpolation formula* (3.18) *to obtain*

$$\mathfrak{L}f = \sum_1^n B_i f(x_i) + R(f), \quad \text{where} \quad B_i = \mathfrak{L}L_i(x). \tag{4.5}$$

Let us look at the case of the functionals (4.2). By Theorem 2 we know that the best integration formula is

$$\int_a^b f(x)\,dx = \sum_1^n B_i f(x_i) + R_I(f), \quad \text{where} \quad B_i = \int_a^b L_i(x)\,dx, \tag{4.6}$$

that the best differentiation formula is

$$f'(x) = \sum_1^n f(x_i) L_i'(x) + R_D(f), \tag{4.7}$$

and that the best interpolation formula is

$$f(x) = \sum_1^n f(x_i) L_i(x) + R_f(x). \tag{4.8}$$

This last approximation is perhaps the most remarkable. Of course, operating with the interpolation operator on both sides of (3.18) does not change anything. However, it proves

COROLLARY 2. *The natural spline interpolation formula* (3.18) *is also the best interpolation formula of order m in the sense of Sard.*

Although the formulae (3.18) and (4.8) are identical, the formula (4.8) represents the result of a totally different procedure: Assume x to be fixed; among the equations

$$f(x) = \sum_1^n B_i f(x_i) + \int_a^b K(x,t) f^{(m)}(t)\,dt \tag{4.9}$$

which are exact if $f \in \pi_{m-1}$, we pick the one which is the solution of the problem

$$\int_a^b \{K(x,t)\}^2\,dt = \text{minimum}.$$

The result is that in (4.9) we have $B_i = L_i(x)$, giving (4.8).

In brief, here is the difference between the old and the new in ENA.

The old: *We have an approximation formula* (2.3) *for* $\mathfrak{L}f$ *in the sense of Newton, provided that*

$$Rf = 0 \quad \textit{whenever} \quad f \in \pi_{n-1}. \tag{4.10}$$

The new: *We have a best approximation formula* (4.5) *for* $\mathfrak{L}f$ *of order* m, *in the sense of Sard, provided that*

$$R(f) = 0 \quad \textit{whenever} \quad f \in \mathcal{S}_{2m-1}(x_1,\ldots,x_n). \tag{4.11}$$

Let me close this chapter with two remarks.

1. *Sard's procedure reduces to Newton's if* $m=n$. I first mention that Theorem 1 is still valid if $m=n$. However, in this case we have a Lagrange interpolant of the data (3.12), call it $P_{m-1}(x)$, which is a polynomial of degree $m-1$. The inclusion

$$P_{m-1}(x) \in \mathcal{S}_{2m-1}(x_1,\ldots,x_n),$$

and the unicity in Theorem 1, show that $P_{m-1}(x)$ is identical with the spline interpolant $S(x)$ of Theorem 1. Thus, in this case, when $m=n$, we have

$$\pi_{n-1} = \mathcal{S}_{2n-1}(x_1,\ldots,x_n),$$

and natural spline interpolation is identical with π_{n-1} interpolation of the data. It clearly follows that the two methods of approximation of functionals are also identical.

2. *An extremum property of the natural spline interpolant* $S(x)$. We state this as

THEOREM 3. *Let* $m<n$ *and let* $S(x)$ *denote the spline interpolant* (3.13) *of Theorem* 1. *Let* $f(x)$ *be any function interpolating our data, hence*

$$f(x_i) = y_i \qquad (i=1,\ldots,n),$$

and such that $f(x)$ *has* $m-1$ *continuous derivatives, and even an* mth *derivative* $f^{(m)}(x)$ *such that the integral*

$$\int_a^b \left(f^{(m)}(x)\right)^2 dx \tag{4.12}$$

is finite. Then

$$\int_a^b \left(f^{(m)}\right)^2 dx \geqslant \int_a^b \left(S^{(m)}\right)^2 dx,$$

with equality only if $f(x)=S(x)$ *for* $a\leqslant x\leqslant b$.

This shows that $S(x)$ may be characterized among all interpolants $f(x)$ by the condition of giving the integral (4.12) its least value.

As stated in our Introduction the reader will find proofs for all theorems stated above in Greville's paper [2].

I conclude by quoting a comment by Philip Davis made in 1964: "Spline approximation contains the delicious paradox of Prokofieff's Classical Symphony: it seems as though it might have been written several centuries ago, but of course it could not have been."

Problems

1. Determine explicitly the natural spline interpolation formula (3.18) for the order $m=1$, and the knots $x_1=0, x_2=1,\ldots,x_n=n-1$. Use it to show that

$$\int_0^{n-1} f(x)\,dx = \frac{1}{2}f(0) + \sum_{i=1}^{n-2} f(i) + \frac{1}{2}f(n-1) + R_f$$

is the best quadrature formula (4.6) for $m=1$.

2. We refer to §3 and the N.S.I.F. (3.18). Let $n=3$ with $x_1=-1$, $x_2=0$, $x_3=1$, and select $m=2$. Find explicitly the expressions for the three fundamental function $L_i(x)\quad(i=1,2,3)$ in (3.18), by using their definition (3.15).

3. For the knots of Problem 2, $(x_1=-1,\ x_2=0,\ x_3=1)$, and $m=2$, find the coefficients of the best quadrature formula (4.6)

$$\int_{-1}^{1} f(x)\,dx = B_1 f(-1) + B_2 f(0) + B_3 f(1) + R(f).$$

It has the property that $R(f)=0$ if $f \in \mathbb{S}_3(-1,0,1)$.

References

[1] P. J. Davis, *Interpolation and Approximation*, Blaisdell, New York, 1963.

[2] T. N. E. Greville, *Introduction to Spline Functions*, in Theory and Applications of Spline Functions, T. N. E. Greville, ed., Academic Press, New York, 1969, pp. 1–35.

[3] Arthur Sard, *Linear Approximation*, American Mathematical Society, Providence, R.I., 1963.

[4] I. J. Schoenberg, *On Best Approximations of Linear Operators*, Indag. math., 67 (1964) 155–163.

[5] Four lectures given by I. J. Schoenberg at Swarthmore College in the fall of 1966 under an IBM Grant for the Development of Mathematics, ed. by Steven Maurer, Mimeographed Notes, Swarthmore College, Swarthmore, Pa.

CHAPTER **11**

ON PEANO CURVES AND THEIR NON-DIFFERENTIABILITY

1. Introduction. We deal with the subject of a short note of mine which is over forty years old [**4**]. Very recently James Alsina, a 1978 graduate of Middlebury College, made an interesting contribution to the subject (Theorem 4 below, see [**1**]). Its derivation, by an approach differing from Alsina's, is the main content of this chapter.

Let me say a few words on the topics that appear here, the reader being asked to skip them if they seem familiar. These topics are: (i) *Binary and ternary infinite expansion of reals*, (ii) *Cantor's middle-third set* Γ, (iii) *area-filling curves*, also called *Peano curves*, (iv) *continuous functions having no derivatives*.

(i) Representations of *integers* n to the base 2 are sometimes useful in the proof of a theorem depending on n, if it can first be established when n is a power of 2. Digital computers handle numbers in their binary representations. In fact we used the binary representation in the equation (4.6) of MTE 3 dealing with Fibonacci and Lucas numbers. The corresponding *infinite expansions* of reals, while of no possible direct use in Numerical Analysis, were much used by G. Cantor, the founder of set theory. Every real $t \in I = \{0 \leqslant t \leqslant 1\}$ admits a representation

$$t = \frac{a_0}{3} + \frac{a_1}{3^2} + \cdots + \frac{a_n}{3^n} + \cdots = 0_3.a_0a_1a_2\ldots, \tag{1.1}$$

where the digits a_n assume one of the three values $0, 1, 2$. (1.1) is unique, except for the ambivalence shown by

$$0_3.a_0a_1\ldots a_{n-1}1000\ldots = 0_3.a_0a_1\ldots a_{n-1}0222\ldots. \tag{1.2}$$

(ii) The famous and important middle-third set Γ of Cantor (see W. Rudin's book [**3**, pp. 34–35]) may be defined as the set of reals t, such that in their representation (1.1) the digits a_n assume only the values 0 or 2. It is therefore the set

$$\Gamma = \left\{ t;\, t = \frac{2a_0}{3} + \frac{2a_1}{3^2} + \cdots + \frac{2a_n}{3^n} + \cdots, \quad \text{where} \quad a_n = 0 \text{ or } 1 \right\}. \tag{1.3}$$

Observe that this representation of t is unique, because the first representation (1.2) may not occur, as it contains the digit 1.

(iii) Let $x(t)$ and $y(t)$ be real-valued continuous functions in $I=[0,1]$. Late into the 19th century the equations

$$x = x(t), \quad y = y(t) \qquad (0 \leqslant t \leqslant 1) \tag{1.4}$$

were thought to define a curve. Great was the surprise when in 1890 G. Peano exhibited an example of (1.4) where *the curve passed through every point of the unit square* $I^2=\{0\leqslant x, y\leqslant 1\}$. Peano's example was later simplified by H. Lebesgue [**2**, pp. 44–45] and again later by the author in [**4**].

(iv) The first example of a function $F(t)$, which has nowhere a finite derivative $F'(t)$, is due to Weierstrass. For a fine discussion of Weierstrass's $F(t)$, see [**3**, pp. 125–127]. Our derivation of Alsina's results (Theorem 4) is based on a modification of Weierstrass's example (Theorem 3).

2. An Identity on the Cantor Set Γ. Throughout this chapter $f(t)$ denotes the function defined as follows. We first define it in $[0,1]$ by

$$f(t) = \begin{cases} 0 & \text{if } 0\leqslant t\leqslant \frac{1}{3} \\ 3t-1 & \text{if } \frac{1}{3}\leqslant t\leqslant \frac{2}{3} \\ 1 & \text{if } \frac{2}{3}\leqslant t\leqslant 1, \end{cases} \tag{2.1}$$

and extend its definition to all real t such that $f(t)$ is an *even* function which is *periodic of period* 2. This means that we have

$$f(-t) = f(t), \quad f(t+2) = f(t) \quad \text{for all } t. \tag{2.2}$$

Its graph is shown in Figure 11.1.

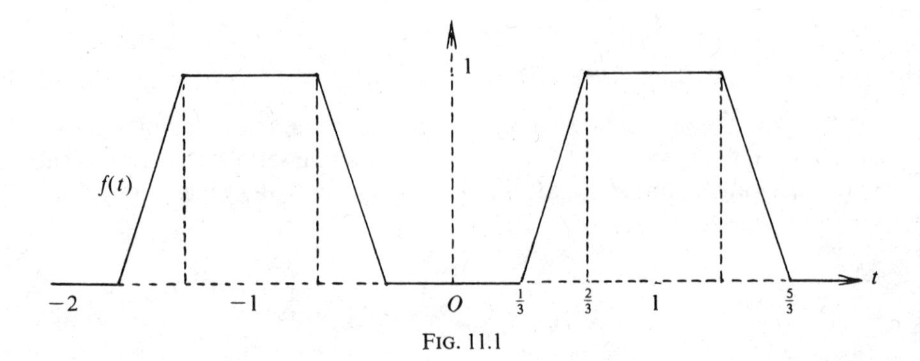

FIG. 11.1

The main property of this function is that it produces the following remarkable identity on Γ (see also Problem 4 at the end of this chapter).

LEMMA 1. *We have*

$$t=\sum_{0}^{\infty}\frac{2f(3^n t)}{3^{n+1}}\quad \textit{if}\quad t\in\Gamma. \tag{2.3}$$

Proof: If

$$t=\sum_{0}^{\infty}\frac{2a_n}{3^{n+1}}\qquad (a_n=0,1), \tag{2.4}$$

then (2.3) would follow from the relations

$$a_n=f(3^n t)\qquad (n=0,1,\ldots). \tag{2.5}$$

To prove (2.5) observe that (2.4) implies

$$3^n t=3^n\left(\frac{2a_0}{3}+\cdots+\frac{2a_{n-1}}{3^n}\right)+\frac{2a_n}{3}+\frac{2a_{n+1}}{3^2}+\cdots\Big),$$

whence

$$3^n t=N_n+\frac{2a_n}{3}+\frac{2a_{n+1}}{3^2}+\cdots\qquad (N_n \text{ is an even number}). \tag{2.6}$$

From the graph of $f(t)$ we conclude the following:

$$\text{If } a_n=0,\quad \text{then } N_n\leqslant 3^n t\leqslant N_n+\frac{2}{3^2}+\frac{2}{3^3}+\cdots=N_n+\frac{1}{3}$$

and therefore $f(3^n t)=0$.

$$\text{If } a_n=1,\quad \text{then } N_n+\frac{2}{3}\leqslant 3^n t\leqslant N_n+\frac{2}{3}+\frac{2}{3^2}+\cdots=N_n+1$$

and so $f(3^n t)=1$.

This establishes (2.5) and the relation (2.3).

3. Two Peano Curves. The plane curve is defined by the equations

$$x(t)=\sum_{0}^{\infty}\frac{1}{2^{n+1}}f(3^{2n}t), \tag{3.1}$$

$$y(t)=\sum_{0}^{\infty}\frac{1}{2^{n+1}}f(3^{2n+1}t)\qquad (0\leqslant t\leqslant 1). \tag{3.2}$$

If $t \in \Gamma$, hence

$$t = \sum_0^\infty \frac{2a_n}{3^{n+1}} \qquad (a_n = 0, 1), \tag{3.3}$$

by (2.5) we may write (3.1) and (3.2) as

$$x(t) = \sum_0^\infty \frac{1}{2^{n+1}} a_{2n}, \quad y(t) = \sum_0^\infty \frac{1}{2^{n+1}} a_{2n+1}. \tag{3.4}$$

We may now invert these relationships: Let

$$P = (x(t), y(t))$$

be an *arbitrarily preassigned point of the square* $I^2 = \{0 \leqslant x, y \leqslant 1\}$, and regard (3.4) as the binary expansions of the coordinates of P. This defines a_{2n} and a_{2n+1}, and therefore also the full sequence a_n. With it we define t $(\in \Gamma)$ by (3.3), and now (3.4), being a consequence of (3.1) and (3.2), shows that the point P is on our curve. This proves

THEOREM 1. *The mapping*

$$t \to (x(t), y(t)) \tag{3.5}$$

from I *into* I^2, *defined by* (3.1), (3.2), *covers the square* I^2 *even if we restrict* t *by*

$$t \in \Gamma. \tag{3.6}$$

This result extends naturally to higher dimensions. We discuss only the case of the space curve

$$X(t) = \sum_0^\infty \frac{1}{2^{n+1}} f(3^{3n} t), \tag{3.7}$$

$$Y(t) = \sum_0^\infty \frac{1}{2^{n+1}} f(3^{3n+1} t), \tag{3.8}$$

$$Z(t) = \sum_0^\infty \frac{1}{2^{n+1}} f(3^{3n+2} t), \qquad (0 \leqslant t \leqslant 1). \tag{3.9}$$

Indeed, if we define t by (3.3), then again (2.5) shows that

$$X(t) = \sum_0^\infty \frac{1}{2^{n+1}} a_{3n}, \quad Y(t) = \sum_0^\infty \frac{1}{2^{n+1}} a_{3n+1}, \quad Z(t) = \sum_0^\infty \frac{1}{2^{n+1}} a_{3n+2}. \tag{3.10}$$

If the right-hand sides are the binary expansions of the coordinates of an arbitrarily preassigned point of I^3, then this point of I^3 is reached by our space curve for the value of $t \in \Gamma$ defined by (3.3). This proves

THEOREM 2. *The mapping*

$$t \to (X(t), \quad Y(t), \quad Z(t)) \tag{3.11}$$

from I into the cube I^3, defined by (3.7), (3.8), (3.9), *covers I^3, even if we restrict t by*

$$t \in \Gamma. \tag{3.12}$$

An important question: *Are our mappings* (3.5) *and* (3.11) *continuous*? That they are continuous follows immediately by Weierstrass's M-test, because our expansions (3.1), (3.2), (3.7), (3.8), (3.9) are *termwise dominated* by the series of constants

$$1 = \sum_{0}^{\infty} \frac{1}{2^{n+1}}.$$

This ensures their *uniform convergence* and therefore also the *continuity* of their sums.

4. A Modification of Weierstrass's Example. The very first nowhere differentiable function is given by

Weierstrass's Example. Let $0<a<1$, and let b be an odd integer such that

$$ab > 1 + \frac{3\pi}{2} = 5.712. \tag{4.1}$$

Then the function

$$W(t) = \sum_{0}^{\infty} a^n \cos(b^n \pi t) \tag{4.2}$$

is nowhere differentiable.

This we modify by replacing in (4.2) the function $\cos \pi t$ by the linear Euler spline $E(t)$ that interpolates $\cos \pi t$ at all integer values of t. Its graph is shown in Figure 11.2. Evidently $E(t)$ is linear between any two consecutive integers and $E(n) = (-1)^n$ for all integers n.

Our aim is to establish

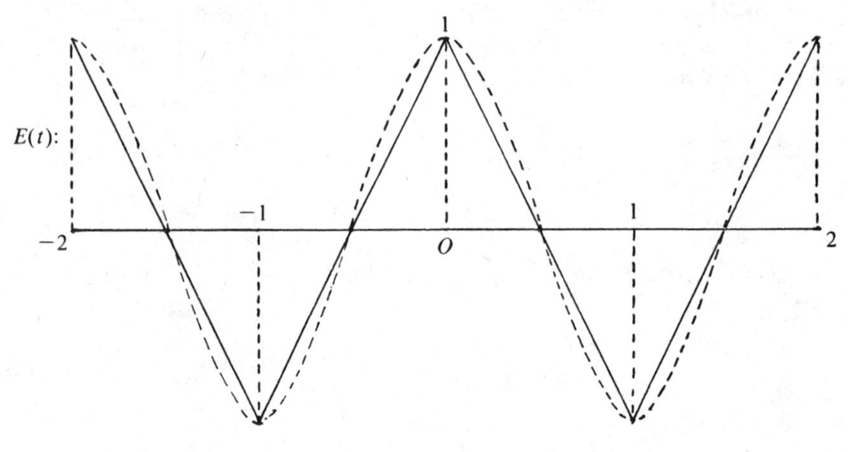

FIG. 11.2

THEOREM 3. *Let* $0<a<1$, *and let* b *be an odd integer such that*

$$ab > 4. \tag{4.3}$$

Then the function

$$F(t) = \sum_0^\infty a^n E(b^n t) \tag{4.4}$$

is nowhere differentiable.

A proof only requires an adaptation of the skillful version of Weierstrass's proof as given by Rudin in [**3**, pp. 125–127]. This adaptation is as follows.
We define

$$s_m(t) = \sum_0^{m-1} a^n E(b^n t), \quad r_m(t) = \sum_m^\infty a^n E(b^n t) \tag{4.5}$$

and the corresponding difference quotients

$$\phi_m(h) = \frac{s_m(t+h) - s_m(t)}{h}, \quad \psi_m(h) = \frac{r_m(t+h) - r_m(t)}{h}. \tag{4.6}$$

Since $E(t)$ has slopes ± 2, we have by (4.5)

$$s'_m(t) = \sum_0^{m-1} a^n b^n 2\epsilon_n(t), \quad \text{where } \epsilon_n(t) = \pm 1,$$

and therefore

$$|s_m'(t)| \leqslant 2\sum_0^{m-1} a^n b^n = 2\frac{(ab)^m-1}{ab-1}.$$

By (4.6) this shows that

$$|\phi_m(h)| = \left|\frac{1}{h}\int_t^{t+h} s_m'(u)\,du\right| \leqslant 2\frac{(ab)^m-1}{ab-1} < 2\frac{(ab)^m}{ab-1}. \tag{4.7}$$

Next, given t and m, there is an integer k_m such that

$$|b^m t - k_m| \leqslant \tfrac{1}{2}. \tag{4.8}$$

Defining

$$q_m = b^m t - k_m, \tag{4.9}$$

we choose the increment

$$h_m = \frac{1-q_m}{b^m}. \tag{4.10}$$

By (4.8) and (4.9) we have $|q_m| \leqslant \frac{1}{2}$, hence $\frac{1}{2} \leqslant 1-q_m \leqslant \frac{3}{2}$, and therefore

$$0 < h_m \leqslant \frac{3}{2b^m}. \tag{4.11}$$

Now (4.9) and (4.10) show that

$$t + h_m = \frac{b^m t + 1 - q_m}{b^m} = \frac{1+k_m}{b^m},$$

and so

$$b^n(t+h_m) = b^{n-m}(1+k_m). \tag{4.12}$$

Let us assume that $n \geqslant m$. Since b is an odd integer, we conclude that $b^{n-m}k_m$ has the same parity as k_m. From (4.12) we conclude that

$$E\big(b^n(t+h_m)\big) = E\big(b^{n-m} + b^{n-m}k_m\big) = E(1+k_m)$$

and so

$$E\big(b^n(t+h_m)\big) = (-1)^{k_m+1}. \tag{4.13}$$

From Figure 11.2 we see that $E(t)$ is odd about the point $t=\frac{1}{2}$. This means that $E(1-t)=-E(t)$, or $E(1+t)=-E(-t)=-E(t)$. Hence $E(t+1)=-E(t)$, and iterating we obtain that

$$E(t+k)=(-1)^k E(t).$$

From (4.9), and for $n \geqslant m$, we obtain that

$$\begin{aligned} E(b^n t) &= E(b^{n-m} b^m t) = E(b^{n-m}(k_m + q_m)) \\ &= E(b^{n-m} k_m + b^{n-m} q_m) = E(b^{n-m} q_m + k_m) \end{aligned}$$

and finally that

$$E(b^n t) = (-1)^{k_m} E(b^{n-m} q_m). \tag{4.14}$$

By (4.6), (4.13) and (4.14) we find that

$$\begin{aligned} \psi_m(h_m) &= \frac{1}{h_m} \sum_{n=m}^{\infty} a^n \{E(b^n(t+h_m)) - E(b^n t)\} \\ &= \frac{1}{h_m} \sum_{n=m}^{\infty} a^n \{(-1)^{k_m+1} + (-1)^{k_m+1} E(b^{n-m} q_m)\} \end{aligned}$$

and finally that

$$\psi_m(h_m) = \frac{(-1)^{k_m+1}}{h_m} \sum_{n=m}^{\infty} a^n \{1 + E(b^{n-m} q_m)\}.$$

Here all terms of the series are nonnegative; taking only the first term for $n=m$, we obtain

$$|\psi_m(h_m)| \geqslant \frac{a^m}{h_m}(1 + E(q_m)).$$

From $|q_m| \leqslant \frac{1}{2}$ we have $E(q_m) \geqslant 0$, and (4.11) shows that

$$|\psi_m(h_m)| \geqslant \frac{a^m}{h_m} \geqslant \frac{2}{3}(ab)^m. \tag{4.15}$$

Now (4.7) and (4.15) allow us to conclude that

$$\begin{aligned} \left| \frac{F(t+h_m) - F(t)}{h_m} \right| &= |\phi_m(h_m) + \psi_m(h_m)| \\ &\geqslant |\psi_m(h_m)| - |\phi_m(h_m)| \\ &\geqslant \frac{2}{3}(ab)^m - 2\frac{(ab)^m}{ab-1} \\ &= 2(ab)^m \left(\frac{1}{3} - \frac{1}{ab-1} \right). \end{aligned}$$

If $ab-1>3$, hence if (4.3) holds, the last lower bound $\to\infty$ as $m\to\infty$ and proves our theorem.

REMARKS. On comparing the two lower bounds (4.1) and (4.3) we see that the number π of (4.1) has been replaced by the smaller number 2. Observe that π and 2 are the maximal slopes of the functions $\cos\pi t$ and $E(t)$, respectively, and this is the origin of these numbers. This decrease of the bound from 5.712 to 4 will make it possible to apply Theorem 3 to obtain Alsina's Theorem 4, which is our main result. This is done in our next and last section.

5. The Nondifferentiability of the Peano Curves. Concerning the two Peano curves

$$C_2:\begin{cases} x(t)=\sum_0^\infty \frac{1}{2^{n+1}} f(3^{2n}t) \\ y(t)=\sum_0^\infty \frac{1}{2^{n+1}} f(3^{2n+1}t) \qquad (0\leqslant t\leqslant 1), \end{cases} \tag{5.1}$$

and

$$C_3:\begin{cases} X(t)=\sum_0^\infty \frac{1}{2^{n+1}} f(3^{3n}t) \\ Y(t)=\sum_0^\infty \frac{1}{2^{n+1}} f(3^{3n+1}t) \\ Z(t)=\sum_0^\infty \frac{1}{2^{n+1}} f(3^{3n+2}t) \qquad (0\leqslant t\leqslant 1), \end{cases} \tag{5.2}$$

we wish to establish the following main result.

THEOREM 4 (James Alsina). *The curves C_2 and C_3 are nowhere differentiable. This means that there is no value of t such that the velocity vector $(x'(t), y'(t))$ exists and is finite, and a similar statement concerns $(X'(t), Y'(t), Z'(t))$.*

A proof requires a second linear Euler spline $S(t)$ defined by

$$S(t) = \tfrac{3}{4}(E(t-1)+1), \tag{5.3}$$

where $E(t)$ is the function used in §4. Like $E(t)$, $S(t)$ is a continuous even function of period 2; its graph in the interval $[0,2]$ is the tall isosceles triangle of Figure 11.3. Notice that the lower zig-zag polygon, continued

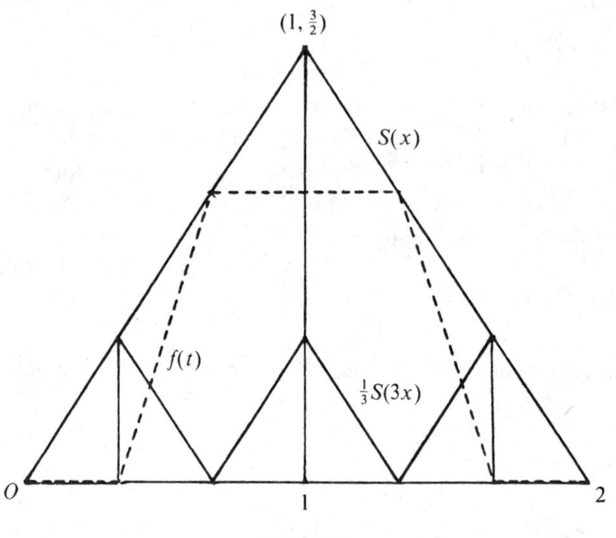

FIG. 11.3

with period 2, represents the function $\frac{1}{3}S(3t)$. Also that if we subtract the lower graph from the upper graph we get in $[0,2]$ the graph of our function $f(t)$ of Figure 11.1. Since all these functions have the same period 2, we have established the identity

$$f(t) = S(t) - \tfrac{1}{3}S(3t) \quad \text{for all real } t. \tag{5.4}$$

A proof that the curve C_2 is nowhere differentiable. From (5.1) and (5.4) we obtain

$$2x(t) = \sum_0^\infty \frac{1}{2^n} f(3^{2n}t) = \sum_0^\infty \frac{1}{2^n} S(3^{2n}t) - \frac{1}{3}\sum_0^\infty \frac{1}{2^n} S(3^{2n+1}t). \tag{5.5}$$

Likewise we have

$$2y(t) = \sum_0^\infty \frac{1}{2^n} f(3^{2n+1}t) = \sum_0^\infty \frac{1}{2^n} S(3^{2n+1}t) - \frac{2}{3}\sum_0^\infty \frac{1}{2^{n+1}} S(3^{2n+2}t).$$

If in the last series we write $n+1=n'$, then n' runs from 1 to ∞. However, we may let n' run from 0 to ∞, provided that we subtract out the new term thus introduced. This gives

$$2y(t) = \sum_0^\infty \frac{1}{2^n} S(3^{2n+1}t) - \frac{2}{3}\sum_0^\infty \frac{1}{2^n} S(3^{2n}t) + \frac{2}{3}S(t). \tag{5.6}$$

If we multiply the equation (5.5) by 3 and add it to (5.6), then one of the series drops out and we obtain the identity

$$6x(t)+2y(t) = \left(3-\frac{2}{3}\right)\sum_0^\infty \frac{1}{2^n} S(3^{2n}t) + \frac{2}{3}S(t),$$

and finally

$$3x(t)+y(t) = \frac{7}{6}\sum_0^\infty \frac{1}{2^n} S(9^n t) + \frac{1}{3}S(t). \tag{5.7}$$

If in (4.4) we choose $a=\frac{1}{2}$ and $b=9$, which is an odd integer, we find that $ab=9/2=4.5$, and so the inequality (4.3) is satisfied, and by Theorem 3 the function

$$F(t) = \sum_0^\infty \frac{1}{2^n}E(9^n t) \quad \textit{is nowhere differentiable}. \tag{5.8}$$

On the other hand (5.3) shows that

$$S(9^n t) = \frac{3}{4}E(9^n t-1)+\frac{3}{4} = \frac{3}{4}E(9^n(t-1))+\frac{3}{4},$$

the last expression being correct because $9^n \equiv 1 \pmod 2$. Therefore, by (5.3) we obtain that

$$\sum_0^\infty \frac{1}{2^n}S(9^n t) = \frac{3}{4}F(t-1)+\frac{3}{2},$$

and (5.8) shows that *the right-hand side of* (5.7) *is nowhere differentiable*. This proves Theorem 4 for C_2: If $x'(t_0)$ and $y'(t_0)$ would exist and be finite, then (5.7) would imply that the right-hand side is differentiable for $t=t_0$.

Returning to (5.7) the reader might have possible doubts for integer values of t, where $S(t)$ has corners. However, from the proof of Theorem 3 we know that the *one-sided derivative* $F'_+(t)$, of $F(t)$, cannot exist and be finite, and the same must be true for the function $3x(t)+y(t)$ of (5.7).

A kinematic interpretation: We have shown that the rectilinear motion obtained by orthogonal projection of the motion (5.1) onto the line $y-3x=0$ never has a finite velocity. Zeno would have puzzled about this.

The case of C_3 follows along similar lines, only a little more complicated. From (5.2) and (5.4) we have

$$\begin{aligned}
2X(t) &= \sum_0^\infty \frac{f(3^{3n}t)}{2^n} = \sum_0^\infty \frac{S(3^{3n}t)}{2^n} - \frac{1}{3}\sum_0^\infty \frac{S(3^{3n+1}t)}{2^n}\\
2Y(t) &= \sum_0^\infty \frac{f(3^{3n+1}t)}{2^n} = \sum_0^\infty \frac{S(3^{3n+1}t)}{2^n} - \frac{1}{3}\sum_0^\infty \frac{S(3^{3n+2}t)}{2^n}\\
2Z(t) &= \sum_0^\infty \frac{f(3^{3n+2}t)}{2^n} = \sum_0^\infty \frac{S(3^{3n+2}t)}{2^n} - \frac{1}{3}\sum_0^\infty \frac{S(3^{3n+3}t)}{2^n},
\end{aligned} \tag{5.9}$$

where the very last series may be written as

$$-\frac{2}{3}\sum_{0}^{\infty}\frac{S(3^{3n+3}t)}{2^{n+1}}=-\frac{2}{3}\sum_{0}^{\infty}\frac{S(3^{3n}t)}{2^{n}}+\frac{2}{3}S(t).$$

If we now multiply the equations (5.9) respectively by $3, 1, \frac{1}{3}$, and add them together, we obtain the identity

$$6X(t)+2Y(t)+\frac{2}{3}Z(t)=\frac{25}{9}\sum_{0}^{\infty}\frac{S(3^{3n}t)}{2^{n}}+\frac{2}{9}S(t). \qquad (5.10)$$

We realize that here we have to apply Theorem 3 for $a=\frac{1}{2}$ and $b=3^{3}=27$, and the remainder of the proof is so much like the previous case of C_2, that we may omit details.

REMARKS. 1. Most mathematical problems are solved up to a point, but not completely, and our problem is no exception. In Problem 2, at the end of this chapter, the reader is asked to show that the functions $x(t),\dots,Z(t)$, are nondifferentiable on the Cantor set Γ. The same question on the complement $I\setminus\Gamma$ remains unanswered.

2. Inspecting the equations (5.1) and (5.2) we realize that there should also be a "one-dimensional Peano curve" C_1, and that its equation should be

$$C_1\colon \xi(t)=\sum_{0}^{\infty}\frac{1}{2^{n+1}}f(3^{n}t) \qquad (0\leqslant t\leqslant 1). \qquad (5.11)$$

While the curve C_2, or C_3, must work hard to cover the square I^2, or the cube I^3, notice that C_1 covers I *effortlessly*, by its mere continuity, because $\xi(0)=0$ and $\xi(1)=1$. Even so it is easy to show that $\xi(t)$ *is nondifferentiable* on Γ. But what about its complement $I\setminus\Gamma$?

Also $\xi(t)$ may be expressed in terms of our $S(t)$, and we find that

$$\xi(t)=\frac{1}{6}\sum_{0}^{\infty}\frac{1}{2^{n}}S(3^{n}t)+\frac{1}{3}S(t). \qquad (5.12)$$

Presumably the sum of the series on the right-hand side is nowhere differentiable, but Theorem 3 does not apply to prove it, because if $a=\frac{1}{2}$ and $b=3$, then $ab=3/2<4$. In contrast to this, that

$$\sum_{0}^{\infty}\frac{1}{2^{n}}S(2^{n}y) \text{ is nowhere differentiable}$$

is shown simply and elegantly in R. P. Boas, *A primer of real variables*, Carus Monograph No. 13, 1960, pp. 115–116.

Problems

1. Show that in the notations of §3 we have

$$x(\tfrac{1}{2}) = y(\tfrac{1}{2}) = X(\tfrac{1}{2}) = Y(\tfrac{1}{2}) = Z(\tfrac{1}{2}) = \tfrac{1}{2}.$$

Hint: Determine for $t = \frac{1}{2}$ the value of $3^n t$ modulo 2.

2. Show that the function

$$x(t) = \sum_0^\infty \frac{1}{2^{n+1}} f(3^{2n} t)$$

is nondifferentiable for all t in the Cantor set Γ. The same is true for the functions $y(t), X(t), Y(t), Z(t)$.

Hint: Use the fact established by (3.3) and (3.4): On Γ the function $x(t)$ may be defined by the mapping

$$t = \sum_0^\infty \frac{2a_n}{3^{n+1}} \rightarrow x(t) = \sum_0^\infty \frac{1}{2^{n+1}} a_{2n}.$$

If $a_{2m} = 0$, choose the increment $\delta_m = 2/3^{2m+1}$. If $a_{2m} = 1$, choose the increment $\delta_m = -2/3^{2m+1}$. In both cases study behavior of $(x(t+\delta_m) - x(t))/\delta_m$ as $m \rightarrow \infty$.

3. Use the ambivalence expressed by the equation (1.2) to show that the mapping (3.5), of Theorem 1, maps four distinct points of I onto the point

$$x_0 = 0_2.101, \quad y_0 = 0_2.111$$

of I^2. This happens whenever the coordinates x_0 and y_0 have terminating binary expansions.

4. Show that the function $S(t)$, defined by (5.3), or Figure 11.3, admits the expansion

$$S(t) = \sum_0^\infty \frac{f(3^n t)}{3^n} \quad \text{for all real } t.$$

Restricted to $0 \leqslant t \leqslant 1$, this expansion shows that in Lemma 1 we may omit the assumption that $t \in \Gamma$ and replace it by $t \in I$.

Hint: Solve (5.4) for $S(x)$ by iterating the identity (5.4) indefinitely.

5. Verify the identity $\sin^4 t = \sin^2 t - \frac{1}{4}\sin^2 2t$ and use it to show that

$$\sum_0^{\infty} \frac{\sin^4(2^n t)}{4^n} = \sin^2 t.$$

Hint: Proceed as in Problem 4.

References

[1] James Alsina, *The Peano curve of Schoenberg is nowhere differentiable*, J. Approx. Theory, 33 (1981) 28–42.

[2] Henri Lebesgue, *Leçons sur l'intégration et la recherche des fonctions primitives*, Gauthier-Villars, Paris, 1928.

[3] Walter Rudin, *Principles of Mathematical Analysis*, McGraw-Hill, New York, 1953.

[4] I. J. Schoenberg, *On the Peano curve of Lebesgue*, Bull. Amer. Math. Soc., 44 (1938) 519.

CHAPTER **12**

ON THE ARITHMETIC-GEOMETRIC MEAN AND SIMILAR ITERATIVE ALGORITHMS

1. Introduction. We deal with the following problem belonging to the interface between analysis and numerical analysis. Let the continuous functions $f(x, y)$ and $g(x, y)$ be positive and explicitly defined for all positive values of x and y. We define

$$x_1 = f(x, y), \quad y_1 = g(x, y), \tag{1.1}$$

and iterate these relations by means of the algorithm

$$x_n = f(x_{n-1}, y_{n-1}), \qquad y_n = g(x_{n-1}, y_{n-1}) \qquad (n=1,2,\ldots), \tag{1.2}$$

where we write $x_0 = x$, $y_0 = y$. The problem is to discover cases when *the sequences x_n and y_n converge to the same limit, i.e.,*

$$\lim x_n = \lim y_n = \Phi(x, y) \tag{1.3}$$

and the limit function $\Phi(x, y)$ can be determined explicitly.

This problem is reminiscent of the problem of finding an explicit expression for the general integral of an ordinary first-order differential equation. There are two differences between these problems: 1. Our present problem is more difficult and can be solved in fewer special cases. 2. Fortunately, while our problem is fascinating, it can certainly not be compared in importance with the problem of differential equations.

The problem of the *existence* of solutions of differential equations has been investigated since the time of Cauchy. A similar theory is lacking for our present problem. In this matter of *existence*, the order of difficulty of the two problems is reversed: Our problem is much easier. However, since the equations (1.1) represent the most general continuous mapping of the positive quadrant $Q = \{x > 0, y > 0\}$ into itself, it is clear that very strong assumptions are necessary to ensure the existence of $\Phi(x, y)$. In §2 we present sufficient conditions to ensure the existence of $\Phi(x, y)$ and allow us to describe its nature.

We also discuss two famous algorithms: The *arithmetic-geometric mean of Gauss* (§3) and the *Borchardt algorithm* (§§4 and 5). Browsing in the library of the Military Academy at West Point, I found a reference to the long-forgotten work [**5**] of J. Schwab, published in Nancy in 1813, in which the author develops Borchardt's algorithm long before Borchardt was born, by a geometric approach of great beauty. To appreciate Schwab's contribution we must realize that he never heard of Gauss's Theorem 3, established in 1799, but not published until 1818. I am grateful to my friend Pierre-Jean Laurent, of Grenoble, for sending me, after a long search, a photocopy of Schwab's book from the old Royal Library in Paris.

The reader should regard the present chapter as an introduction to B. C. Carlson's paper [**1**], where important new contributions to our problem can be found. We propose one of Carlson's new algorithms as a problem at the end of the chapter. Further references are John Todd's paper [**6**] and F. G. Tricomi's paper [**7**]. See these papers also for valuable historical information.

2. On the Convergence of Iterative Algorithms. We assume that the functions $f(x, y)$ and $g(x, y)$ of our Introduction are *means* and that they are *comparable*. Here are the definitions.

DEFINITION 1. *We say that $f(x, y)$ is a mean, and write $f \in M$, provided that $f(x, y)$ is defined and continuous in the domain*

$$Q' = \{0 < x \leq y\} \tag{2.1}$$

and that it satisfies the following two conditions:

1. *Property B of betweenness*:

$$\text{If} \quad x < y \quad \text{then} \quad x < f(x, y) < y. \tag{2.2}$$

This already implies by continuity that

$$\text{if} \quad x = y \quad \text{then} \quad x = f(x, y) = y, \tag{2.3}$$

hence that

$$f(x, x) = x \quad \text{if} \quad x > 0. \tag{2.4}$$

2. *Property H of homogeneity of degree one*:

$$f(\lambda x, \lambda y) = \lambda f(x, y) \quad \text{if} \quad \lambda > 0. \tag{2.5}$$

DEFINITION 2. *Let $f \in M$ and $g \in M$, both having Property B. We say that the means f and g are comparable and write $f \prec g$, provided that*

$$f(x, y) \leqslant g(x, y), \quad \textit{with equality if and only if} \quad x = y. \tag{2.6}$$

Notice that the "if" part of (2.6) is a priori true, for $x = y$ implies $f(x, x) = x = g(x, x)$ by (2.4).

Sometimes we replace (2.1) by the entire quadrant

$$Q = \{x > 0, y > 0\}. \tag{2.7}$$

This is surely permissible in the case that f and g are *symmetric*; hence $f(x, y) = f(y, x)$, $g(x, y) = g(y, x)$. Occasionally we may replace Q' by

$$Q'' = \{0 < y \leqslant x\}, \tag{2.8}$$

in which case (2.2) is to be replaced by

$$\textit{if} \quad x > y \quad \textit{then} \quad x > f(x, y) > y. \tag{2.9}$$

Examples of means and of pairs of comparable means are numerous. The simplest are

$$A = A(x, y) = \frac{x + y}{2}, \quad \text{the } \textit{arithmetic} \text{ mean}, \tag{2.10}$$

$$G = G(x, y) = \sqrt{xy}, \quad \text{the } \textit{geometric} \text{ mean}, \tag{2.11}$$

$$H = H(x, y) = \frac{2}{\frac{1}{x} + \frac{1}{y}} = \frac{2xy}{x + y}, \quad \text{the } \textit{harmonic} \text{ mean}, \tag{2.12}$$

$$E = E(x, y) = \sqrt{\frac{x^2 + y^2}{2}}, \quad \text{the } \textit{euclidean} \text{ mean}. \tag{2.13}$$

Figure 12.1 shows the geometric construction of these means. It shows at a glance that A, G, H, and E are indeed means in the sense of our Definition 1. Also that

$$H \prec G, \quad G \prec A, \quad A \prec E. \tag{2.14}$$

Notice that the points O and H are *harmonic conjugates* with respect to the points x and y, because of the geometric relation $HA \cdot OA = (xA)^2$, and this is why H is called the *harmonic* mean.

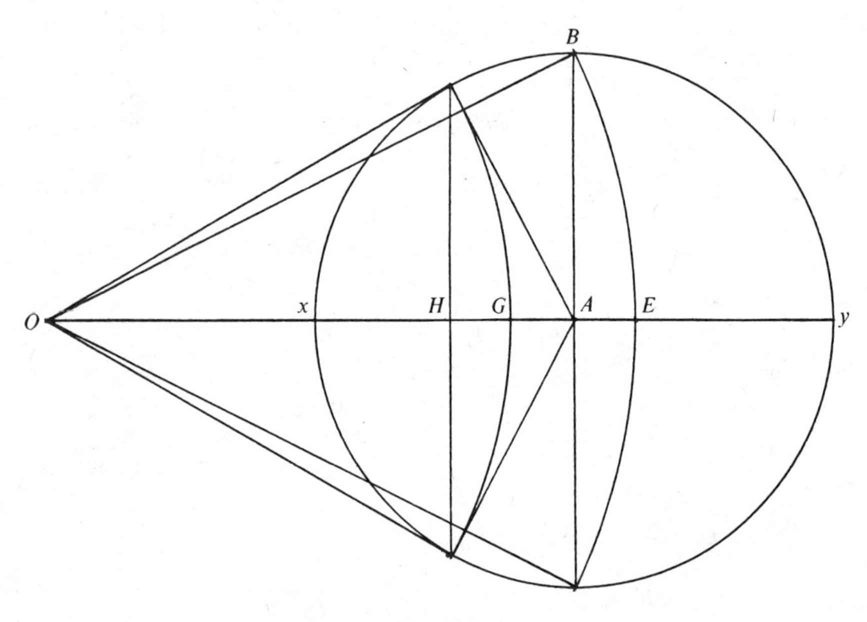

FIG. 12.1

An important generalization of these are the *power means*

$$M_r(x, y) = \left(\frac{x^r + y^r}{2}\right)^{1/r} \qquad (r \neq 0). \tag{2.15}$$

Evidently

$$A = M_1, \quad H = M_{-1}, \quad E = M_2.$$

That M_r has Property B is clear from the inequalities

$$x < \left(\frac{x^r + y^r}{2}\right)^{1/r} < y \quad \text{if} \quad x < y. \tag{2.16}$$

In Kazarinoff's book [**3**, pp. 63–65] the following results are established, even for n variables rather than two:

$$M_0(x, y) = \lim_{r \to 0} M_r(x, y) = G(x, y), \tag{2.17}$$

$$\lim_{r \to -\infty} M_r(x, y) = x, \quad \lim_{r \to \infty} M_r(x, y) = y. \tag{2.18}$$

As r increases between $-\infty$ and $+\infty$, $M_r(x, y)$ increases from x to y. (2.19)

It follows that

$$\textit{if} \quad r < s, \quad \textit{then} \quad M_r(x, y) < M_s(x, y). \tag{2.20}$$

The main result of this section is

THEOREM 1. *Let f and g be comparable means in Q'; hence* (2.6) *holds in the domain* (2.1). *Then the algorithm*

$$x_n = f(x_{n-1}, y_{n-1}), \quad y_n = g(x_{n-1}, y_{n-1}) \qquad (n=1,2,\ldots; x_0=x, y_0=y) \tag{2.21}$$

has the property that

$$\lim x_n = \lim y_n = \Phi(x, y), \tag{2.22}$$

where $\Phi(x, y)$ is a mean in the domain Q'.

In §3 we will show that, if $f=G$ and $g=A$, then the common limit $\Phi(x, y)=\mu(x, y)$ can be expressed in terms of an elliptic integral. The function $\mu(x, y)$ is usually called the *arithmetic geometric mean*, which we abbreviate to

$$\mu(x, y) = (G, A) \text{ mean.}$$

Accordingly we may call $\Phi(x, y)$, of (2.22), the (f, g) mean, and write

$$\Phi = (f, g) \text{ mean.} \tag{2.23}$$

Proof of Theorem 1. If $x=y$, then by (2.21) and (2.4) we have $x_n=y_n=x=y$ for all n and (2.22) holds trivially with

$$\Phi(x, x) = x. \tag{2.24}$$

Assume now that $x<y$. Combining (2.2) with (2.6) we obtain

$$x < f(x, y) < g(x, y) < y, \tag{2.25}$$

and so $x<x_1<y_1<y$. This shows that the algorithm (2.21) can be continued indefinitely leading to

$$x < x_1 < \cdots < x_n < y_n < \cdots < y_1 < y \quad \text{for all } n. \tag{2.26}$$

These inequalities ensure the existence of the limits

$$\lim x_n = \xi, \quad \lim y_n = \eta, \quad \text{with } \xi \leqslant \eta. \tag{2.27}$$

Letting $n\to\infty$ in (2.21), the continuity of f and g shows that

$$\xi = f(\xi,\eta), \quad \eta = g(\xi,\eta). \tag{2.28}$$

Let us show that

$$\xi = \eta. \tag{2.29}$$

This is immediate, because the opposite assumption that $\xi<\eta$ would imply, by (2.6), that

$$\xi = f(\xi,\eta) < g(\xi,\eta) = \eta,$$

which contradicts the property (2.2) of both means f and g. This establishes the relations (2.22).

There remains to show that

$$\Phi(x, y) \quad \textit{is a mean}, \tag{2.30}$$

hence that Φ satisfies the conditions of Definition 1. This we do in stages.

1. *The inequalities* (2.2) follow directly from (2.26) and (2.22), which imply that $x<\Phi(x, y)<y$.

2. *That Φ is homogeneous of degree* 1 we see as follows. The functions $x_1=f(x, y)$, $y_1=g(x, y)$ are homogeneous by assumption. But then also $x_2=f(x_1, y_1)$, $y_2=g(x_1, y_1)$, regarded as functions of x, y, also have this property, because

$$f(f(\lambda x,\lambda y), g(\lambda x,\lambda y))$$
$$= f(\lambda f(x, y),\lambda g(x, y)) = \lambda f(f(x, y), g(x, y)).$$

By induction it follows that

$$x_n = f_n(x, y), \quad y_n = g_n(x, y) \quad \text{are homogeneous of degree 1.} \tag{2.31}$$

But then

$$\Phi(\lambda x,\lambda y) = \lim f_n(\lambda x,\lambda y) = \lambda \lim f_n(x, y) = \lambda\Phi(x, y),$$

which completes the proof.

There remains the last property

3. $\Phi(x, y)$ *is continuous in the domain Q'*. Let us assume that $x<y$. Besides

$$x < x_1 < \cdots < x_n < \Phi(x, y) < y_n < \cdots < y_1 < y, \tag{2.32}$$

using obvious notations, we consider the inequalities

$$x' < x_1' < \cdots < x_n' < \Phi(x', y') < \cdots < y_1' < y'. \tag{2.33}$$

Let ϵ be positive, and let (x', y') satisfy

$$|x - x'| < \epsilon, \quad |y - y'| < \epsilon. \tag{2.34}$$

We are to show that *for a preassigned* $\delta > 0$ *we have*

$$|\Phi(x, y) - \Phi(x', y')| < \delta, \tag{2.35}$$

provided that ϵ *is sufficiently small.*

Given δ, we choose n, in (2.32), so that

$$y_n - x_n < \delta/2. \tag{2.36}$$

We observe next that $x_n = f_n(x, y)$ and $y_n = g_n(x, y)$ are continuous functions. We can therefore choose ϵ, in (2.34), so small that

$$|x_n - x_n'| < \delta/2, \quad |y_n - y_n'| < \delta/2. \tag{2.37}$$

We claim that (2.35) holds. Indeed, by (2.36) and (2.37), we have

$$\begin{aligned}\Phi(x, y) - \Phi(x', y') &< y_n - x_n' = y_n - x_n + (x_n - x_n')\\ &< \delta/2 + \delta/2 = \delta,\end{aligned}$$

and

$$\begin{aligned}\Phi(x, y) - \Phi(x', y') &> x_n - y_n' = x_n - y_n + (y_n - y_n')\\ &> -(\delta/2) - (\delta/2) = -\delta,\end{aligned}$$

and therefore (2.35) holds. This completes a proof of Theorem 1.

The invariance property of $\Phi(x, y)$. Suppose that we start the algorithm (2.21) with $n = 2$ rather than $n = 1$. This means that we start it from (x_1, y_1) rather than (x, y). Since all later (x_n, y_n) are unchanged, we must evidently have

$$\Phi(x, y) = \Phi(x_1, y_1). \tag{2.38}$$

This proves

COROLLARY 1. *The function* $\Phi(x, y)$ *of Theorem* 1 *satisfies in* Q' *the functional equation*

$$\Phi(x, y) = \Phi(f(x, y), g(x, y)). \tag{2.39}$$

We call (2.38) the *invariance property* of Φ, and $\Phi(x, y)$ is called the *invariant function of the algorithm* (2.21). Its fundamental role for our problem is shown by

THEOREM 2. *Let $\Psi(x, y)$ be a mean, hence $\Psi \in M$, having the invariance property*

$$\Psi(x, y) = \Psi(x_1, y_1). \tag{2.40}$$

Then

$$\Psi(x, y) = \Phi(x, y), \tag{2.41}$$

so that Ψ coincides with the limit function of Theorem 1.

Proof: From (2.40) by iteration we get the string of equations

$$\Psi(x, y) = \Psi(x_1, y_1) = \Psi(x_2, y_2) = \cdots = \Psi(x_n, y_n).$$

This and (2.22) show that

$$\Psi(x, y) = \lim \Psi(x_n, y_n) = \Psi(\lim x_n, \lim y_n) = \Psi(\Phi, \Phi) = \Phi(x, y) \tag{2.42}$$

by the property (2.4) of the mean Ψ. This completes our proof.

We have also established

COROLLARY 2. *The function $\Phi(x, y)$ of Theorem* 1 *is the unique mean satisfying the functional equation* (2.39).

REMARKS. 1. Observe that in proving Theorem 2 we have not used all properties of the mean Ψ. In fact we have used only the following three properties: (1) the invariance (2.40), (2) the continuity of Ψ, and (3) that $\Psi(x, x) = x$. The remaining properties of Ψ follow, of course, from these, because of (2.41), but need not be assumed to start with.

2. *Theorem 2 is the main tool in verifying that an explicitly given $\Phi(x, y)$ is the (f, g) mean of x and y.* It will be applied in proving Gauss's Theorem 3, and also in §5 to establish Schwab's Theorem 4.

We conclude this section with two examples. In view of Theorem 1 and (2.20) we may consider the (M_r, M_s) mean, if $r < s$. Recalling that $H = M_{-1}$ and $G = M_0$, the choice $r = -1$, $s = 1$, gives us

EXAMPLE 1. Let

$$x_1 = \frac{2xy}{x + y}, \quad y_1 = \frac{x + y}{2}. \tag{2.43}$$

Notice that $x_1 y_1 = xy$, which shows the invariance property of the product

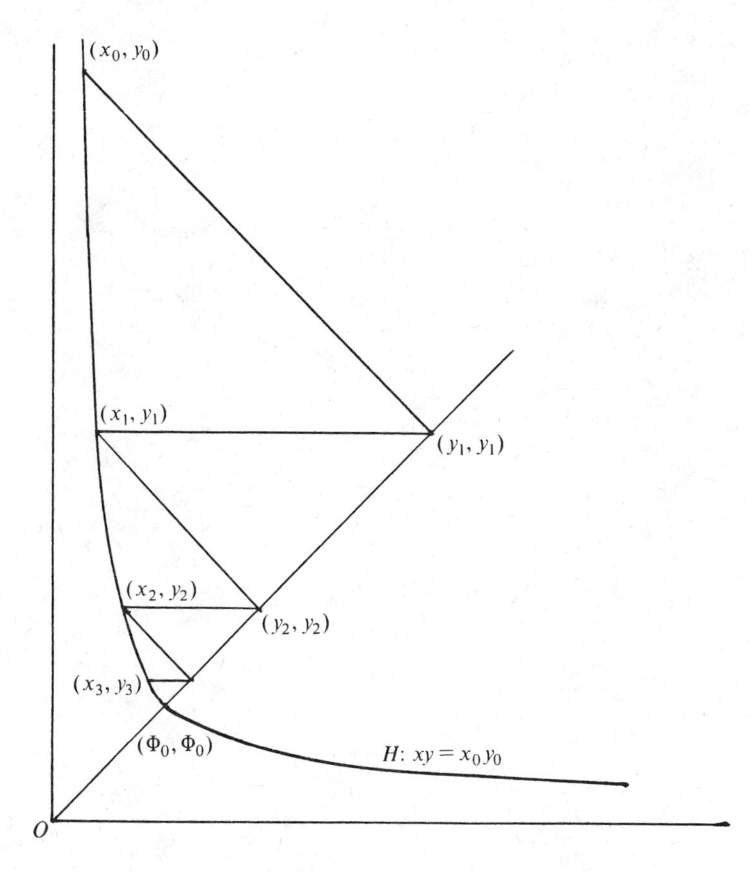

FIG. 12.2

xy. As we need the homogeneity of degree 1 we conclude by Theorem 2 that the limit function is

$$\Phi(x, y) = \sqrt{xy}\,. \tag{2.44}$$

Let $x_0 = x$, $y_0 = y$. Figure 12.2 shows the hyperbola $H: xy = x_0 y_0$ passing through (x_0, y_0). From (x_0, y_0) we drop a perpendicular to the line $y = x$, intersecting it in the point (y_1, y_1) (why?). The horizontal line $y = y_1$ intersects H in (x_1, y_1), because all points (x_n, y_n) are on H (why?). We iterate and get the zig-zag line which shows that (x_n, y_n) converges fast to the vertex (Φ_0, Φ_0) of H, where $\Phi_0 = \sqrt{x_0 y_0}$.

EXAMPLE 2. Here we invert the process; we start from a given mean Φ and look for appropriate means f and g such that $\Phi = (f, g)$ mean. Let

$$\Phi(x, y) = \sqrt[3]{xy^2}\,, \tag{2.45}$$

which is certainly a mean. Again let $x_0 = x$, $y_0 = y$, and Figure 12.3 shows the curve

$$\Gamma: xy^2 = x_0 y_0^2$$

intersecting the line $y = x$ in the point (Φ_0, Φ_0), where $\Phi_0 = \sqrt[3]{x_0 y_0^2}$. At this point Γ has a tangent having the slope $-1/2$. Imitating Figure 12.2, we draw through (x_0, y_0) a line also of slope $-1/2$, hence having the equation

$$x + 2y = x_0 + 2y_0,$$

and therefore intersecting $y = x$ in the point (y_1, y_1), where

$$y_1 = \frac{x_0 + 2y_0}{3}.$$

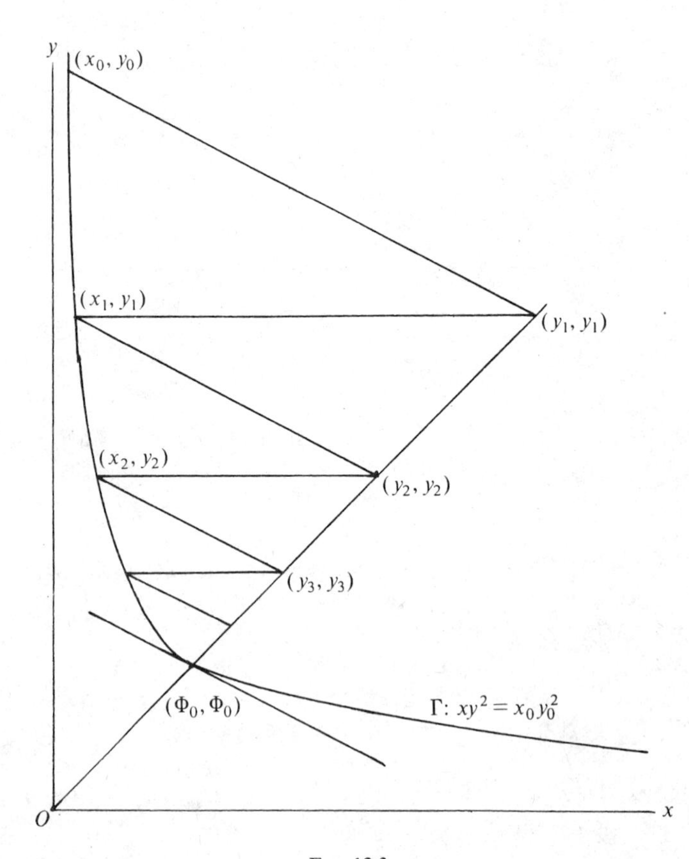

FIG. 12.3

From the invariance property $x_1 y_1^2 = x_0 y_0^2$ we find that

$$x_1 = \frac{x_0 y_0^2}{y_1^2} = \frac{9x_0 y_0^2}{(x_0+2y_0)^2}.$$

These equations show that we should iterate the equations

$$x_1 = \frac{9xy^2}{(x+2y)^2}, \quad y_1 = \frac{x+2y}{3}. \tag{2.46}$$

As in Example 1, the fact that the slanting lines are parallel to the tangent to Γ at its vertex ensures the so-called *quadratic* convergence of the process. Starting the algorithm with the initial values $x_0 = a$, $y_0 = 1$, $0 < a < 1$, the iteration of (2.46) will result in

$$\lim x_n = \lim y_n = \sqrt[3]{a}.$$

For $a = .5$ we find, to six decimal places, that $x_4 = y_4 = .793700$.

3. The Arithmetic-Geometric Mean. This is the (A, G) algorithm valid in the entire quadrant $Q = \{x > 0,\ y > 0\}$. For notational reasons we replace (x, y) with (a, b) and assume that $a > b > 0$. Starting from $a = a_0$, $b = b_0$, the algorithm is

$$a_n = \tfrac{1}{2}(a_{n-1} + b_{n-1}), \quad b_n = \sqrt{a_{n-1} b_{n-1}} \qquad (n = 1, 2, \ldots). \tag{3.1}$$

Its remarkable behavior is described by

THEOREM 3 (Gauss). *We have*

$$\lim a_n = \lim b_n = \mu(a, b), \tag{3.2}$$

where

$$\mu(a, b) = 1 \Big/ \frac{2}{\pi} \int_0^{\pi/2} \frac{d\phi}{\sqrt{a^2 \cos^2\phi + b^2 \sin^2\phi}}. \tag{3.3}$$

Using the notation (2.23) we see that $\mu(a, b) = (G, A)$ mean of a and b. From Figure 12.1, or also directly, we find that $b < b_1 < a_1 < a$, with $a_1 - b_1 < \frac{1}{2}(a - b)$, whence $a_n - b_n < (a - b)/2^n$. This ensures the existence and equality of the limits (3.2).

The main point is the *invariance property* of $\Phi(a,b)=\mu(a,b)$ that $\mu(a,b)=\mu(a_1,b_1)$. By (3.3) this amounts to the famous equation

$$\int_0^{\pi/2}\frac{d\phi}{\sqrt{a^2\cos^2\phi+b^2\sin^2\phi}}=\int_0^{\pi/2}\frac{d\psi}{\sqrt{a_1^2\cos^2\psi+b_1^2\sin^2\psi}}. \tag{3.4}$$

The following simple and elegant proof is due to D. J. Newman (see *Lectures on approximation and value distribution* by T. H. Ganelius, W. K. Hayman and D. J. Newman, University of Montreal Press, 1982). We write

$$I(a,b)=\int_0^{\pi/2}\frac{d\phi}{\sqrt{a^2\cos^2\phi+b^2\sin^2\phi}},\ J(a,b)=\int_0^{\infty}\frac{dx}{\sqrt{(x^2+a^2)(x^2+b^2)}}, \tag{3.5}$$

and observe that (3.4) follows from the two equations

$$I(a,b)=J(a,b), \tag{3.6}$$

$$J(a,b)=J(a_1,b_1). \tag{3.7}$$

Indeed, by (3.6) and (3.7) we have

$$I(a,b)=J(a,b)=J(a_1,b_1)=I(a_1,b_1),$$

and so

$$I(a,b)=I(a_1,b_1),$$

proving (3.4).

Proof of (3.6): Changing variables in $I(a,b)$ by setting $x=\tan\phi$, we have the equations

$$dx=\frac{d\phi}{\cos^2\phi},\quad \frac{d\phi}{\cos\phi}=\cos\phi\,dx=\frac{dx}{\sqrt{1+x^2}},$$

hence

$$I(a,b)=\int_0^{\pi/2}\frac{d\phi}{\cos\phi}\frac{1}{\sqrt{a^2+b^2\tan^2\phi}}=\int_0^{\infty}\frac{dx}{\sqrt{(1+x^2)(a^2+b^2x^2)}}$$

and the last integral is equal to $J(a,b)$, as seen if we replace x by x/b.

Proof of (3.7): In

$$J(a_1, b_1) = \frac{1}{2}\int_{-\infty}^{\infty} \frac{dt}{\sqrt{(t^2+a_1^2)(t^2+b_1^2)}}$$

we change variables by setting

$$t = \frac{1}{2}\left(x - \frac{ab}{x}\right),$$

observing that as x increases from 0 to $+\infty$, t increases from $-\infty$ to $+\infty$. From the equations

$$t^2 + a_1^2 = t^2 + \left(\frac{a+b}{2}\right)^2 = \frac{(x^2-ab)^2+(a+b)^2x^2}{4x^2} = \frac{(x^2+a^2)(x^2+b^2)}{4x^2},$$

$$t^2 + b_1^2 = t^2 + ab = \frac{(x^2-ab)^2+4abx^2}{4x^2} = \frac{(x^2+ab)^2}{4x^2},$$

$$dt = \frac{x^2+ab}{2x^2}dx,$$

the integral is seen to become

$$J(a_1, b_1) = \frac{1}{2}\int_0^{\infty} \frac{2x}{\sqrt{(x^2+a^2)(x^2+b^2)}} \cdot \frac{2x}{x^2+ab} \cdot \frac{x^2+ab}{2x^2}dx$$

$$= \int_0^{\infty} \frac{dx}{\sqrt{(x^2+a^2)(x^2+b^2)}} = J(a, b).$$

4. The Schwab-Borchardt Algorithm. This algorithm is a modification of Gauss's algorithm: Starting from positive $x = x_0$, $y = y_0$, we apply the algorithm

$$x_{n+1} = \tfrac{1}{2}(x_n + y_n), \quad y_{n+1} = \sqrt{x_{n+1}y_n} \qquad (n = 0, 1, \ldots). \tag{4.1}$$

Notice that in computing y_{n+1} we are using the "updated" value x_{n+1}.

Is this an algorithm of the form (2.21), *and does our Theorem* 1 *of* §2 *become applicable*?

That the answers are affirmative is seen as follows. That (4.1) is of the form (2.21) is seen if we rewrite (4.1) in the form

$$x_{n+1} = \frac{1}{2}(x_n + y_n), \quad y_{n+1} = \sqrt{\frac{x_n + y_n}{2}y_n}\,. \tag{4.2}$$

Here we already know that $f(x, y)=\frac{1}{2}(x+y)$ is a mean; but also

$$g(x, y) = \sqrt{\frac{x+y}{2}y}$$

is a mean: The continuity and homogeneity are obvious, and also Property B is easily verified, as well as $g \prec f$. We may therefore apply Theorem 1 and conclude that

$$\lim x_n = \lim y_n = \Phi(x, y), \tag{4.3}$$

where $\Phi(x, y)$ is a mean in our sense.

However, the remarkable fact is that the common limit $\Phi(x, y)$ is *explicitly* given by

$$\Phi(x, y) = \begin{cases} \dfrac{\sqrt{y^2-x^2}}{\arccos\dfrac{x}{y}} & \text{if } 0 \leqslant x < y, \\ x & \text{if } x = y, \\ \dfrac{\sqrt{x^2-y^2}}{\operatorname{arccosh}\dfrac{x}{y}} & \text{if } y < x. \end{cases} \tag{4.4}$$

The third case of (4.4), when $y<x$, will be proposed as Problem 5 at the end of the chapter. The first case $0 \leqslant x < y$ is established by the following theorem where, for geometric reasons, we return to the (a, b) notation.

THEOREM 4 (Schwab-Borchardt). *If*

$$0 \leqslant a < b, \tag{4.5}$$

and

$$a_{n+1} = \tfrac{1}{2}(a_n + b_n), \quad b_{n+1} = \sqrt{a_{n+1} b_n} \qquad (n=0,1,\ldots; a=a_0, b=b_0), \tag{4.6}$$

then

$$\lim a_n = \lim b_n = \frac{\sqrt{b^2-a^2}}{\arccos\dfrac{a}{b}}. \tag{4.7}$$

We already know that *the limits exist and are equal* by Theorem I. But also directly it is easily verified that $a<a_1<b_1<b$ and $b_1-a_1<\frac{1}{2}(b-a)$, which imply the italicized statement. There remains to establish the explicit expression (4.7). This is done by Schwab by the following ingenious arguments.

In Figure 12.4 we assume that

$$OA = OC = OB = b, \quad OD = a, \quad AC = CB. \tag{4.8}$$

Let A_1 and B_1 be the midpoints of the segments AC and BC, respectively, and let D and D_1 be the intersections of OC with AB and A_1B_1. Since D_1 is the midpoint of DC, we have

$$OD_1 = \tfrac{1}{2}(OC + OD) = \tfrac{1}{2}(b + a) = a_1. \tag{4.9}$$

Moreover, from the right-angled triangle OA_1C we obtain $(OA_1)^2 = OD_1 \cdot OC$, and so, by (4.8) and (4.9),

$$OA_1 = \sqrt{OD_1 \cdot OC} = \sqrt{a_1 b} = b_1. \tag{4.10}$$

In passing from the triangle OAB to OA_1B_1, we see that the radius $OA = b$ and apothem (or height) $OD = a$ have changed to b_1 and a_1, respectively. Also, $\angle AOB$ and base AB have in OA_1B_1 one-half their former values. On iterating this construction on OAB a total of n times, we obtain a slim triangle OA_nB_n having

$$\begin{aligned} &\text{radius } OA_n = b_n, && \text{height } OD_n = a_n, \\ &\angle A_nOB_n = \frac{1}{2^n}(\angle AOB), && \text{base } A_nB_n = \frac{1}{2^n}AB. \end{aligned} \tag{4.11}$$

If we put together 2^n congruent copies of OA_nB_n, in the shape of a

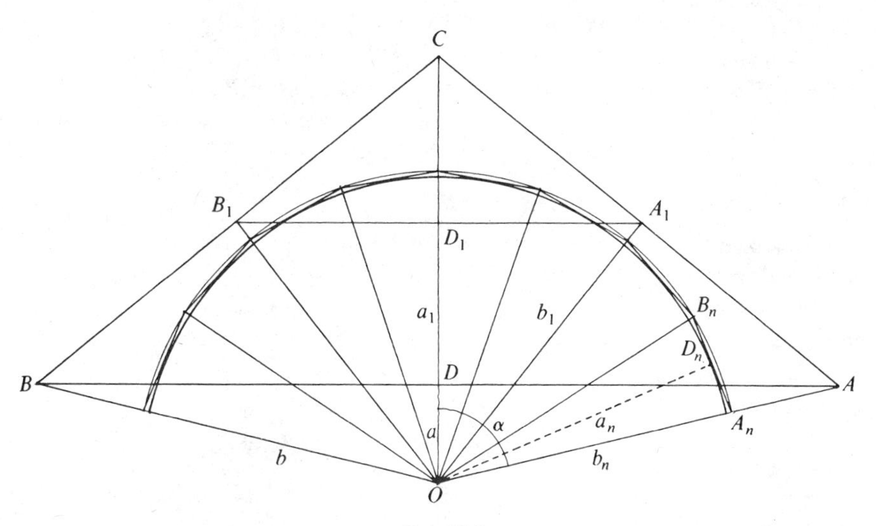

FIG. 12.4

collapsible ladies' fan (Figure 12.4 for $n=3$), we obtain a regular polygonal line Π_n of 2^n sides having, by (4.11),

$$\text{radius} = b_n, \quad \text{apothem} = a_n,$$
$$\text{central angle} = \angle AOB = 2\alpha, \quad \text{perimeter} = AB. \tag{4.12}$$

As $n \to \infty$ it is clear that Π_n converges to a circular arc Γ with central angle $= \angle AOB = 2\alpha$, and length of $\Gamma = AB$. If we denote the radius of Γ by R, it follows that $2\alpha R = AB$, and therefore

$$R = \frac{AB}{2\alpha} = \frac{AD}{\alpha} = \frac{\sqrt{b^2 - a^2}}{\arccos\frac{a}{b}}. \tag{4.13}$$

On the other hand, Figure 12.4 shows that the perimeter of Π_n is contained between the length of the arc of the *inscribed* circle, having radius a_n, and the length of the arc of its *circumscribed* circle, having radius b_n. This means, by (4.11), that $2\alpha a_n < AB = 2\alpha R < 2\alpha b_n$, and therefore

$$a_n < R < b_n.$$

But then letting $n \to \infty$, we obtain

$$\lim a_n = \lim b_n = R. \tag{4.14}$$

Now (4.13) and (4.14) imply the desired result (4.7).

REMARK. The equations (4.9) and (4.10) are the source of Schwab's algorithm, which he states as follows:

Let P be a regular n-gon and let P_1 be a regular $2n$-gon, both having the same perimeter. If a and b are the radii of the inscribed and circumscribed circles of P, and a_1 and b_1 denote the similar quantities of P_1, then

$$a_1 = \frac{a+b}{2}, \quad b_1 = \sqrt{a_1 b}.$$

5. A Second Proof of Theorem 4 by Using Theorem 2. We have already used this "method of the invariant function" proving Gauss's Theorem 3. It may be used whenever an invariant function is explicitly given (see Problems 4 and 5 at the end of this chapter). We are concerned with the algorithm (4.1) and wish to show that

$$\Psi(x, y) = \frac{\sqrt{y^2 - x^2}}{\arccos\frac{x}{y}} \qquad (0 \leqslant x < y) \tag{5.1}$$

satisfies the assumptions of our Theorem 2.

The continuity of Ψ is obvious. Next we wish to show that by extending by continuity to $x=y>0$ we have that

$$\Psi(x,x)=x. \tag{5.2}$$

Keeping y fixed and letting $x\to y$, we have by l'Hôpital's rule

$$\lim_{x\to y}\Psi=\lim\frac{-x}{\sqrt{y^2-x^2}}\Bigg/\left(-\frac{1}{y}\frac{1}{\sqrt{1-\dfrac{x^2}{y^2}}}\right)=\lim x=y=x,$$

which proves (5.2).

There remains to prove the crucial invariance property

$$\frac{\sqrt{y^2-x^2}}{\arccos\dfrac{x}{y}}=\frac{\sqrt{y_1^2-x_1^2}}{\arccos\dfrac{x_1}{y_1}}. \tag{5.3}$$

Notice that

$$y_1^2-x_1^2=\frac{x+y}{2}y-\left(\frac{x+y}{2}\right)^2=\frac{x+y}{2}\left(y-\frac{x+y}{2}\right)=\frac{y^2-x^2}{4}$$

and so $\sqrt{y^2-x^2}=2\sqrt{y_1^2-x_1^2}$, and (5.3) reduces to showing that

$$2\arccos\frac{x_1}{y_1}=\arccos\frac{x}{y}. \tag{5.4}$$

Setting $\alpha=\arccos(x/y)$, or $\cos\alpha=x/y$, dividing (5.4) by 2, we see that (5.4) reduces to

$$\frac{x_1}{y_1}=\cos\frac{\alpha}{2}.$$

But this relation is correct, because

$$\frac{x_1}{y_1}=\frac{x_1}{\sqrt{x_1y}}=\sqrt{\frac{x_1}{y}}=\sqrt{\frac{x+y}{2y}}$$

$$=\sqrt{\frac{(x/y)+1}{2}}=\sqrt{\frac{\cos\alpha+1}{2}}=\cos\frac{\alpha}{2}.$$

Problems

0. The following is merely a numerical exercise with the hand-held calculator rather than a problem: In Theorem 4 of §4 let $a=0$ and $b=1$; use the equations (4.6) and (4.7) to compute π to 6 decimal places by iterating the transformation (4.6) eleven times. Work with 8 decimal places.

1. In Figure 12.1 we have $\angle OAB = 90°$ and $OE = OB$. Verify that the length of the segment OE is equal to the euclidean mean (2.13).

2. Generalize the Example 2 of §2 to approximate the mean

$$\Phi(x, y) = \sqrt[n]{xy^{n-1}} \qquad (0<x\leq y;\ n \text{ is a positive integer}),$$

i.e., find the means f and g such that $\Phi = (f, g)$ mean.

3. Compute the numerical value of the integral

$$\int_0^{\pi/2} \frac{d\phi}{\sqrt{4\cos^2\phi + \sin^2\theta}}$$

by the (A, G) algorithm of Theorem 3.

4. Prove the following

Theorem of Carlson. *For distinct positive $x = x_0$, $y = y_0$, such that $x_0 < y_0$, we perform the algorithm*

$$x_n = \sqrt{x_{n-1}\frac{x_{n-1}+y_{n-1}}{2}}, \quad y_n = \sqrt{\frac{x_{n-1}+y_{n-1}}{2}y_{n-1}} \qquad (n=1,2,\ldots).$$

Show that

$$\lim x_n = \lim y_n = \sqrt{\frac{y^2-x^2}{2\log\frac{y}{x}}}.$$

Hint: Show that the last expression Φ is an invariant function of the algorithm. See [**1**, p. 499].

5. Prove the following

THEOREM OF PFAFF. *If* $y < x$ *in the Schwab-Borchardt algorithm* (4.1), *then*

$$\lim x_n = \lim y_n = \frac{\sqrt{x^2 - y^2}}{\operatorname{arccosh} \frac{x}{y}}.$$

This result was already stated in (4.4).

Hint: Use the method of the invariant function as used in §5. Notice that in $Q'' = \{y \leqslant x\}$ we have $f > g$, and that Theorems 1 and 2 are valid under these new conditions.

References

[1] B. C. Carlson, *Algorithms involving arithmetic and geometric means*, Amer. Math. Monthly, 78 (1971) 496–505.

[2] C. F. Gauss, *Werke*, vol. 3, Göttingen, 1866, pp. 352–355.

[3] N. D. Kazarinoff, *Analytic Inequalities*, Holt, Rinehart and Winston, New York, 1961.

[4] I. J. Schoenberg, *On the arithmetic-geometric mean*, Delta, Univ. of Wisconsin, Madison, 7, no. 2 (Fall 1977) 49–65.

[5] J. Schwab, *Élémens de Géométrie*, Première Partie, Géométrie Plane, Nancy, 1813.

[6] J. Todd, *The lemniscate constants*, Comm. ACM, 18 (1975) 14–19.

[7] F. G. Tricomi, *Sugli algoritmi iterativi nell' analisi numerica*, Accad. Nazionale dei Lincei, 272 (1975) 105–117.

CHAPTER **13**

ON THE KAKEYA-BESICOVITCH PROBLEM

1. Introduction. Few geometrical investigations have aroused as much interest as Besicovitch's solution of the problem proposed by the Japanese mathematician S. Kakeya in 1917. A. S. Besicovitch (1891–1970) was one of the outstanding mathematicians of our century. In 1961 he made a film, "On the problem of Kakeya," for the Mathematical Association of America, which was first shown on May 7, 1962, at the University of Pennsylvania. Kakeya's problem is as follows. Let $\mathrm{U} = AB$ be a unit segment in the plane. We are to move U from its original position AB so as to bring it back to its original position with its endpoints reversed, so that the final position is BA, and during this motion U *should sweep out the least possible area*. Think of U as oozing red paint while it moves: the painted area should be least.

Great was the astonishment when Besicovitch published in 1928 his Theorem 1 of §2 below (see [**1**] and [**2**]). A much simpler proof was given also in 1928 by O. Perron by means of the so-called Perron trees [**4**]. This is an infinite sequence of simple polygons T_m $(m = 0, 1, 2, \ldots)$ of great beauty. Our Figure 13.8 shows the Perron tree T_5. In 1961 the author simplified Perron's construction of his trees and this approach is used in the present chapter [**5**]. The important later work of F. Cunningham will be mentioned in §5.

2. Kakeya's Conjecture and Besicovitch's Theorem. Let us first orient ourselves by a few examples of possible motions.

First example. Figure 13.1 (a) shows the segment AB in its original position. We turn AB around A by 180° obtaining the position AB'. We now slide AB' along its line into the final position BA. Evidently,

$$\text{U } \textit{has swept out a semicircle of area} = \frac{\pi}{2}. \tag{2.1}$$

Second example. Figure 13.1 (b) shows a more economical motion. We rotate AB by 180° around its midpoint 0. This switches the ends and

$$\text{U } \textit{has swept out an area} = \frac{\pi}{4}. \tag{2.2}$$

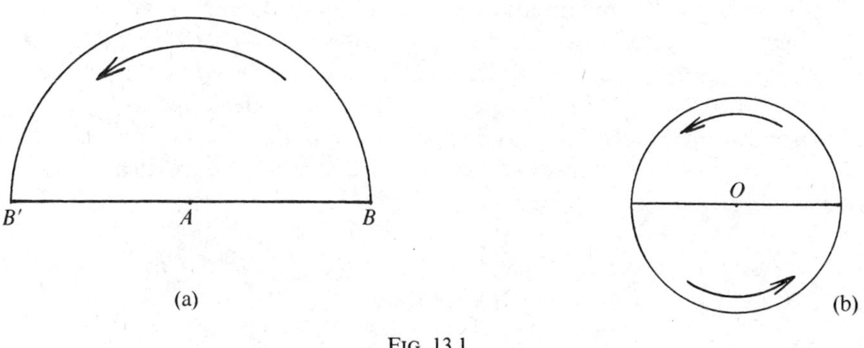

FIG. 13.1

Third example. This example, due to Kakeya, is also historically interesting because it suggested to Kakeya the conjecture below. He considers the three-cusped hypocycloid H of Figure 13.2. This is the curve described by a point on the rim of a circle of radius r that rolls, without slipping, within a fixed circle of radius $R=3r$. We recall the following classical property of this curve: If we draw the tangent to H at a point P on H, and denote by $A'B'$ the portion of the tangent which is in the interior of H, then $A'B'$ is a segment of constant length which is independent of the position of P on H. Let the figure be so scaled that this constant length $A'B'=1$; hence also $AB=1$. If we now start from $P=B$, and let P describe H counterclockwise, we find that the segment $\mathrm{U}=A'B'$ moves inside H and that it returns to the position BA, hence with endpoints reversed, when P reaches B. From calculus we know that

$$\textit{The area of } H=\frac{\pi}{8}. \tag{2.3}$$

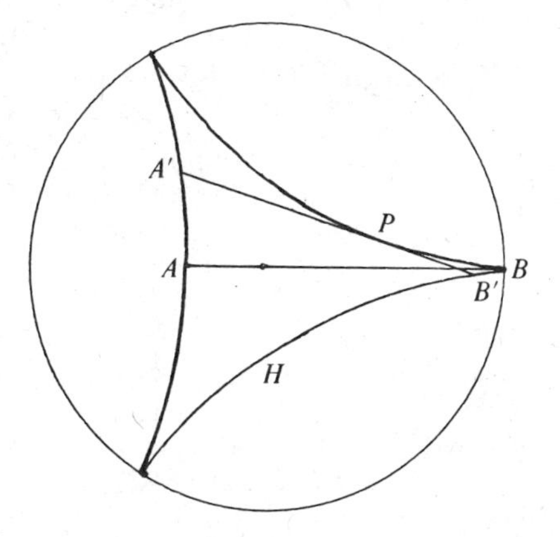

FIG. 13.2

We seem to be doing pretty well, since our examples gave us the decreasing sequence of areas swept out by U as $\pi/2$, $\pi/4$, $\pi/8$. However, a special significance attaches to Kakeya's third example. The fact that in the motion of $A'B'$ of Figure 13.2 there is no freedom whatever *led Kakeya to conjecture that $\pi/8$ is the least value of an area in which* U *can be turned.*

Great was the astonishment of mathematicians when Besicovitch established in 1927 the following

THEOREM 1 (Besicovitch). *The switching of the endpoints of the segment* $U = AB$ *can be done within an arbitrarily small area.**

3. Besicovitch's Lemma and How It Implies a Proof of Theorem 1. Let $T = ABC$ be a triangle of area $=1$, having its base AB on the Ox axis. We divide AB into n equal parts and join them to the vertex C, thereby dividing T into n slim triangles

$$t_1, t_2, \ldots, t_n. \tag{3.1}$$

We translate the triangles t_i, horizontally and independently, to new positions

$$t_1^*, t_2^*, \ldots, t_n^*, \tag{3.2}$$

all still having their bases on Ox. We now define the *Besicovitch constant* B_n by the equation

$$B_n = \min \operatorname{Area}\left(\bigcup_1^n t_i^*\right), \tag{3.3}$$

the minimum being taken with arbitrary translates (3.2). Evidently $B_1 = 1$. Figure 13.3 shows the case when $n = 2$. Although the Besicovitch constant B_n does not depend on the shape of the triangle T of unit area, we have chosen for T a right-angled isosceles triangle. In Problem 1 at the end of this chapter the reader is asked to show that the configuration on the left-side of Figure 13.3 gives the largest possible overlap between t_1^* and t_2^*. The area of $t_1^* \cup t_2^*$ shows that

$$B_2 = \tfrac{2}{3}. \tag{3.4}$$

---

*I have the following story from Besicovitch, who spent a few years in the late 1950's at the University of Pennsylvania, and I think that this is the place to record it: He worked on and solved another essentially equivalent problem in Perm, near the Ural mountains, during the worst time of the Russian civil war, in the winter of 1919. There was no fuel and he worked with his feet in a box filled with straw to keep warm. To the end he did not know if his approach would lead to the desired result. On the strength of this work he was elected in 1920 to a professorship at the University of Leningrad.

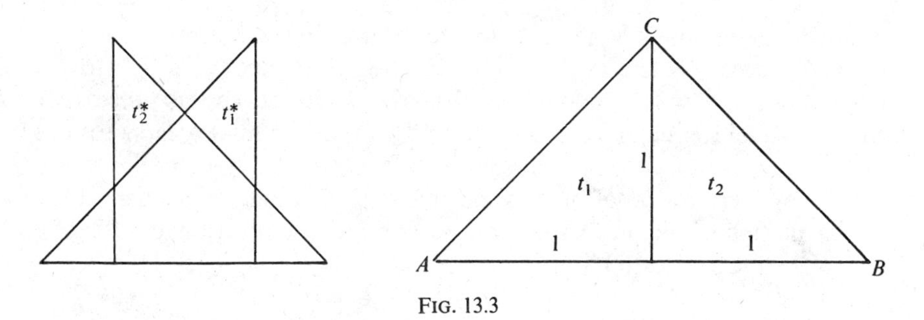

FIG. 13.3

In [**1**] Besicovitch showed by a complicated and difficult proof that

$$\underline{\lim}_{n\to\infty} B_n = 0. \tag{3.5}$$

Restating explicitly this famous result we have the

LEMMA OF BESICOVITCH. *Given $\epsilon > 0$, we can find a value of n such that translating the t_i to appropriate positions t_i^* we have*

$$\text{Area}\left(\bigcup_1^n t_1^*\right) < \epsilon. \tag{3.6}$$

On the principle of "travel now, pay later" we propose to show now how this lemma implies a proof of Theorem 1. We begin with two preliminary observations.

1. *The segment* U$=AB$ *can be moved to a new position A_1B_1 on its line within a zero area.*

Seems obvious, since lines have zero area. We just shift AB along its line from its position AB to A_1B_1.

2. *The unit segment* U$=AB$ *can be moved to any position A_1B_1, which is parallel to AB and has the same length and direction as AB, within an arbitrarily small area.*

Figure 13.4 shows the procedure to be followed. We turn AB to AB' by an arbitrarily small angle δ, so that the line AB' intersects the line A_1B_1 in a

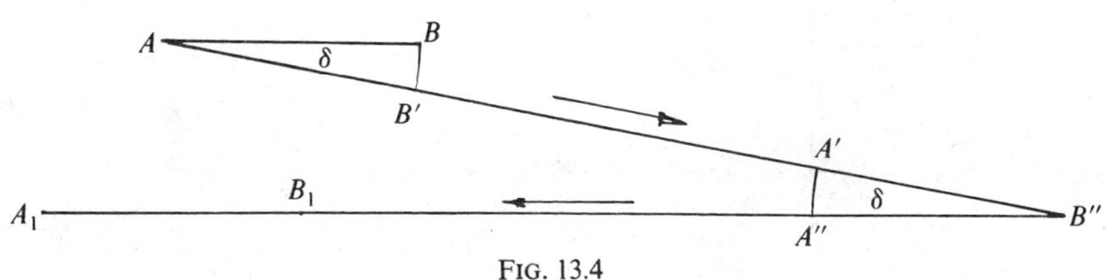

FIG. 13.4

point B''. Now slide AB' along its line to the position $A'B''$. Next turn $A'B''$ to $A''B''$, again by the angle δ. Finally, slide $A''B''$ along its line to the desired position A_1B_1. The only positive areas swept out in this process are the two circular sectors, each of area $\delta/2$, so that the total area swept out is δ.

We now derive a proof of Theorem 1 by an ingenious refinement of the simple motion of Figure 13.1 (a). We start with the southeast half ABD of a square of side 2 (Figure 13.5).

A proof of Theorem 1. We start from the initial position U$=CF$. As in our first example we can reverse U by turning it by 180° to U$'=CF'$ and then shift it along the diagonal AD to the final position FC.

To the triangle ABC we now apply the Lemma of Besicovitch. Given ϵ we can find n such that, if ABC is dissected into the n triangles (3.1), these will by appropriate horizontal shifts assume the positions (3.2) with the property (3.6); hence

$$\text{Area}\left(\bigcup_{1}^{n} t_1^*\right) < \epsilon. \tag{3.7}$$

We repeat this construction for the triangle BDC. Dividing BD into n equal parts, we dissect BDC into n triangles $t_1', t_2', \ldots, t_n'$, with common vertex at C. These we translate vertically to assume positions $t_1'^*, \ldots, t_n'^*$ such that

$$\text{Area}\left(\bigcup_{1}^{n} t_i'^*\right) < \epsilon. \tag{3.8}$$

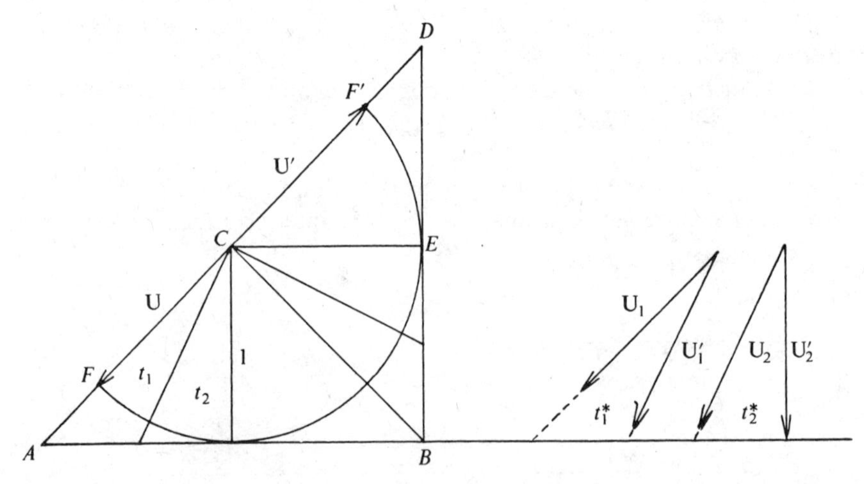

FIG. 13.5

The union of all translated triangles forms the set

$$F = \left(\bigcup_{1}^{n} t_i^* \right) \cup \left(\bigcup_{1}^{n} t_i'^* \right) \tag{3.9}$$

and, by (3.7) and (3.8), we have that

$$\text{Area } F < 2\epsilon. \tag{3.10}$$

Let U_i and U_i' denote unit segments on the left and right lateral sides of t_i^*, both starting from its vertex (see Figure 13.5 showing $\mathrm{U}_1, \mathrm{U}_1', \mathrm{U}_2, \mathrm{U}_2'$). Let likewise U_{n+i} and U_{n+i}' be unit segments on the lower and upper sides of $t_i'^*$. Since

$$\mathrm{U} \| \mathrm{U}_1, \mathrm{U}_1' \| \mathrm{U}_2, \ldots, \mathrm{U}_{2n-1}' \| \mathrm{U}_{2n}, \mathrm{U}_{2n}' \| \mathrm{U}',$$

we can move U to the destination U′ by $2n+1$ parallel displacements as described in Figure 13.4. These displacements require an

$$\text{Area} < (2n+1)\delta.$$

If we select

$$\delta < \frac{\epsilon}{2n+1}, \tag{3.11}$$

we need a total

$$\text{Area} < \epsilon \textit{ for all parallel displacements.} \tag{3.12}$$

As the rotation from U_i to U_i' $(i=1,\ldots,2n)$ takes place within the set F, (3.10) shows that we need an

$$\text{Area} < 2\epsilon \textit{ for all } 2n \textit{ rotations.} \tag{3.13}$$

Therefore the transition from U to U′ requires, by (3.12) and (3.13), an area $< 3\epsilon$, which is arbitrarily small, proving Besicovitch's Theorem 1.

4. And What About the Lemma of Besicovitch? I hope the reader enjoyed the breaking of the triangles of Figure 13.5 into slim triangles and reassembling them to yield a small area. But how is the reassembling to be done? In this we follow Perron [**3**] who discovered the following

THEOREM 2 (O. Perron). *If $n=2^m$, then*

$$B_{2^m} \leqslant \frac{2}{m+2}. \tag{4.1}$$

This is a stronger form of Besicovitch's lemma (3.5). Indeed, (4.1) shows that $B_n \to 0$ if n tends to infinity through the sequence of powers of 2.

Following Perron we establish (4.1) by means of a special sequence of polygons T_m $(m=0,1,2,\ldots)$. T_m has the shape of a tree and this is why I like to call T_m *the mth Perron tree*. However, our construction of T_m is different from Perron's: We let them grow from the ground up, the way trees normally grow [5]. Their growth requires the following

Process of sprouting. We start from a triangle (see Figure 13.6)

$$\tau = ABC \tag{4.2}$$

with its base on the x-axis and having an integer height equal to $k(>1)$. The line $y=k-1$ cuts off from τ the triangle

$$s = EFC$$

which we call *the sprout of* τ. We now let s sprout as follows: Extend the lateral sides of s upwards up to the next integer level $y=k+1$, obtaining the points E' and F'. Finally, join E' to E, and F' to F. Evidently, the lines $E'E$ and $F'F$ are parallel, and they are parallel to the median CD of τ.

> The *sprouting process* is the above procedure whereby the sprout $s=EFC$ gave rise to the heptagon $ABFF'CE'EA$. (4.3)

LEMMA 1. *Draw the triangle $\tau'=A'B'C'$, similar to ABC but having its height $=k+1$, and let τ_1 and τ_2 denote its two halves left and right of its*

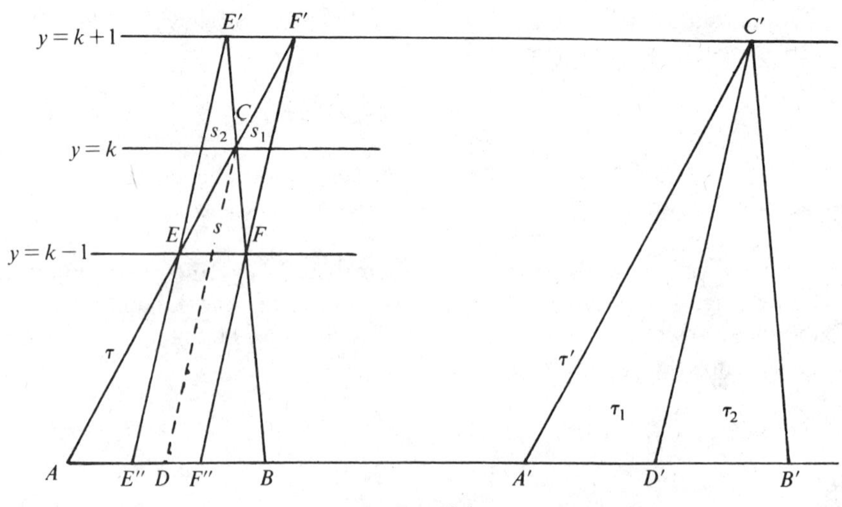

FIG. 13.6

median $C'D'$. Then ($\approx$ means "congruent")

$$\tau_1 \approx AF''F',\ \tau_2 \approx E''BE', \text{ and the left part of Figure 13.6 arises from the right part by translations of } \tau_1 \text{ and } \tau_2. \tag{4.4}$$

Notice that the line $y=k$ cuts off from the heptagon the two sprouts s_1 and s_2 which seem ready to sprout. We note that

$$\text{Area } s = \text{Area } s_1 + \text{Area } s_2, \tag{4.5}$$

a remark to be used below.

We apply the sprouting process to

Growing the Perron trees. Figure 13.7 shows the x-y plane in which we have drawn the level lines

$$y = k \qquad (k=0,1,2,\ldots).$$

We start with the right-angled isosceles triangle of height 2

$$T_0 = ABC, \quad \text{having Area} \quad T_0 = 4. \tag{4.6}$$

This is *the zeroth Perron tree* T_0. We apply to T_0 the sprouting process to obtain *the Perron tree* T_1. It reaches to the level $y=3$, and by (4.5) we find that

$$\text{Area } T_1 = 6. \tag{4.7}$$

It has two sprouts s_1^1 and s_2^1; letting them sprout we obtain T_2 which reaches the level $y=4$. Our Figure 13.8 shows the Perron tree T_5 having $2^5=32$ sprouts. We continue this process until we reach *the tree* T_m with its 2^m vertices on the level $y=m+2$. Some of its properties are as follows.

1. T_m has 2^m sprouts

$$s_1^m, s_2^m, \ldots, s_{2^m}^m, \tag{4.8}$$

which are all within the strip $m+1 \leq y \leq m+2$. It follows from (4.5) that the area of every sprout is passed on to the two sprouts which it generates. It follows that

$$\sum_{i=1}^{2^m} \text{Area } s_i^m = \text{Area } s_1^0 = 1. \tag{4.9}$$

Let us determine the area of T_m. Figure 13.7 shows that

$$\text{Area } T_m \cap \{0 \leq y \leq 1\} = 3, \tag{4.10}$$

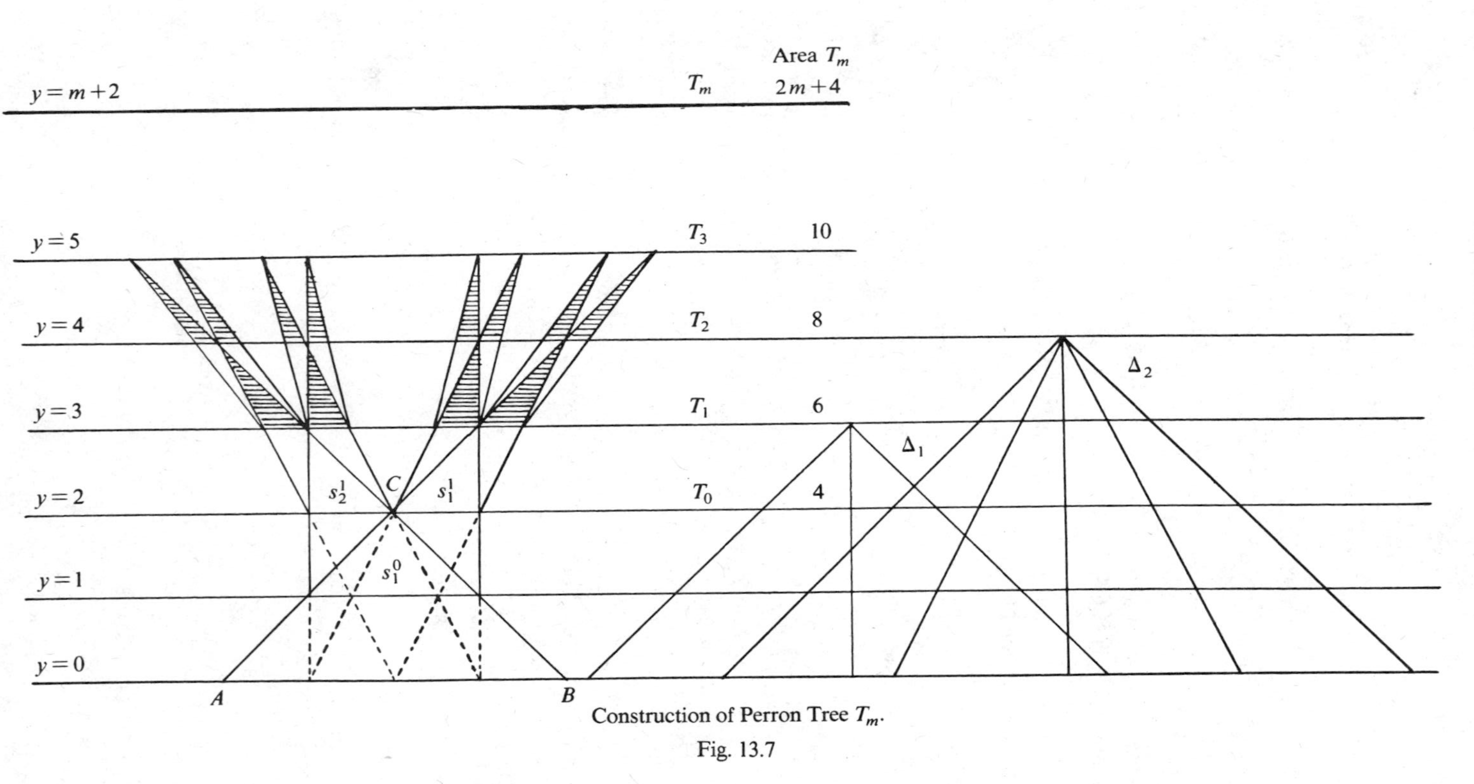

Construction of Perron Tree T_m.

Fig. 13.7

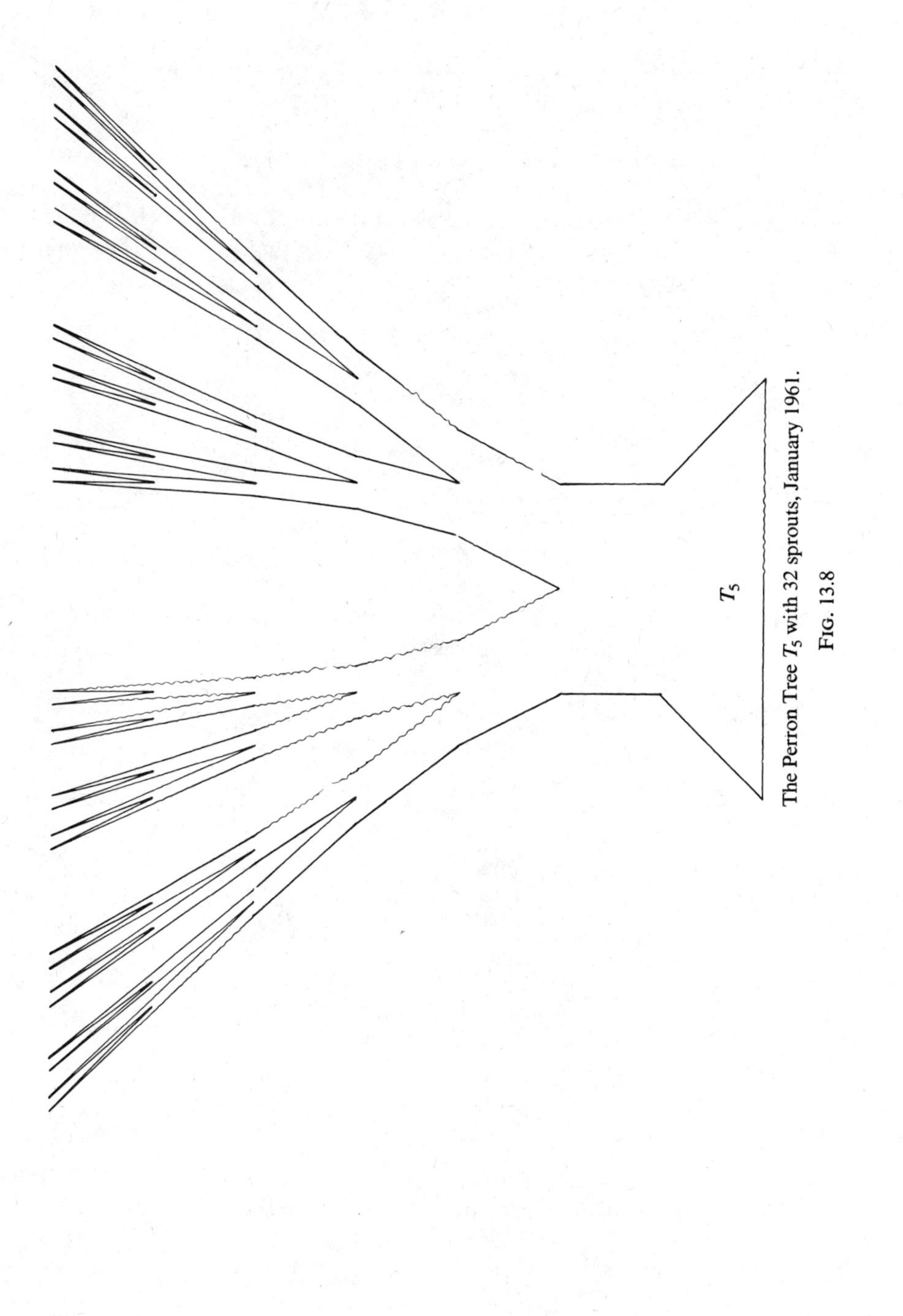

The Perron Tree T_5 with 32 sprouts, January 1961.

FIG. 13.8

while (4.9) may be written as

$$\text{Area } T_m \cap \{m+1 \leqslant y \leqslant m+2\} = 1. \tag{4.11}$$

What about the area A_k of T_m within the strip $\{k \leqslant y \leqslant k+1\}$? Notice inductively that this area

$$A_k = \text{Area } T_m \cap \{k \leqslant y \leqslant k+1\} \qquad (k=1,2,\ldots,m)$$

is composed of 2^{k-1} parallelograms. Passing from A_k to A_{k+1}, each of these parallelograms splits into two parallelograms inheriting its area. Therefore $A_k = A_1 = 2$ and we obtain that

$$\text{Area } T_m \cap \{k \leqslant y \leqslant k+1\} = 2 \qquad (k=1,2,\ldots,m). \tag{4.12}$$

From (4.10), (4.11), and (4.12) we conclude that

$$\text{Area } T_m = 2m+4. \tag{4.13}$$

Also recall that

$$\text{Height of } T_m = m+2. \tag{4.14}$$

2. The essential property of T_m is expressed by the following theorem of Perron from which Theorem 2 will easily follow.

THEOREM 3 (Perron). *Let*

$$\Delta_m = A_m B_m C_m = \bigcup_{i=1}^{2^n} t_i^m \tag{4.15}$$

be a triangle similar to T_0(45°-90°-45° angles), *having the same height as* T_m, *hence* height $\Delta_m = m+2$. *As shown in* (4.15), Δ_m *is divided into* 2^m *slim triangles having equal bases.*

Let

$$s_1^m, s_2^m, \ldots, s_{2^m}^m \tag{4.16}$$

be the sprouts of T_m, *and let*

$$t_1^{m*}, t_2^{m*}, \ldots, t_{2^m}^{m*} \tag{4.17}$$

be the triangles obtained by extending down to $y=0$ *the lateral sides of the sprouts* (4.16) (The (4.16) are the tips of the (4.17).) *Then*

$$t_i^{m*} \approx t_i^m \qquad (i=1,\ldots,2^m), \tag{4.18}$$

and

$$T_m = \bigcup_{1}^{2^m} t_i^{m*}. \tag{4.19}$$

Figure 13.7 shows the triangles Δ_1 and Δ_2 properly dissected.

Proof of Theorem 3. We use induction in m, letting the sprouting process perform the induction step. Assume Theorem 3 to be true for T_{m-1}. Therefore by (4.19)

$$T_{m-1} = \bigcup_{1}^{2^{m-1}} t_i^{(m-1)*}, \tag{4.20}$$

while

$$\Delta_{m-1} = A_{m-1}B_{m-1}C_{m-1} = \bigcup_{1}^{2^{m-1}} t_i^{m-1}, \tag{4.21}$$

and

$$t_i^{(m-1)*} \approx t_i^{m-1} \qquad (i=1,\ldots,2^{m-1}). \tag{4.22}$$

We return to our Figure 13.6 showing the sprouting process. By this process $t_1^{(m-1)*}$ gives rise to the two triangles t_1^{m*} and t_2^{m*}, and generally

$$t_i^{(m-1)*} \text{ generates } t_{2i-1}^{m*} \text{ and } t_{2i}^{m*}. \tag{4.23}$$

This shows that

$$T_m = \bigcup_{i=1}^{2^{m-1}} \left(t_{2i-1}^{m*} \cup t_{2i}^{m*}\right) = \bigcup_{i=1}^{2^m} t_i^{m*},$$

proving (4.19). By our induction assumption we know that

$$t_i^{(m-1)*} \approx t_i^{m-1}, \tag{4.24}$$

while Figure 13.6 shows that appropriate translates of t_{2i-1}^{m*} and t_{2i}^{m*}, we mean t_{2i-1}^{m} and t_{2i}^{m}, fit together to form the triangle

$$t_{2i-1}^{m} \cup t_{2i}^{m}.$$

These are two consecutive triangles of Δ_m proving (4.18).

A proof of Theorem 2 follows readily as follows: Let $\tilde{T}_m$ and $\tilde{\Delta}_m$ be similar copies of (4.19) and (4.15) that are obtained by shrinking T_m and Δ_m in the ratio of $(m+2):1$. Clearly Height $\tilde{T}_m = 1$ and Height $\tilde{\Delta}_m = 1$. By (4.13)

$$\text{Area } \tilde{T}_m = \frac{2m+4}{(m+2)^2} = \frac{2}{m+2}, \quad \text{while Area } \tilde{\Delta}_m = 1. \tag{4.25}$$

From the definition (3.3) of the Besicovitch constant we conclude that

$$B_{2^m} \leqslant \text{Area } \tilde{T}_m = \frac{2}{m+2} \tag{4.26}$$

and Theorem 2 is established.

A perhaps superfluous question: What made the result (4.26) possible? By (4.13) and (4.14) both the area of T_m and the height of T_m grow *linearly* with m. Scaling T_m down to $\tilde{T}_m$ produced (4.25) and (4.26).

5. The Work of F. Cunningham and a Few Conjectures. First a few definitions.

(i) We call the plane set S a *K-set*, provided that within S the endpoints of the unit segment U can be reversed.

(ii) A set S is said to be *star-shaped*, provided that S contains a point O such that every point P of S can be jointed to O by a segment OP which is contained in S.

(iii) We define the constants K, K_1, and K_s as follows:

K = infimum of the areas of K-sets.

K_1 = infimum of the areas of *simply connected* K-sets.

K_s = infimum of the areas of *star-shaped* K-sets.

From 1917 until 1927, when Besicovitch's paper **[1]** appeared, it was thought that

$$K = \frac{\pi}{8} = (.125)\pi. \tag{5.1}$$

This was Kakeya's conjecture. From 1927 to 1965 it was believed that

$$K_1 = \frac{\pi}{8}. \tag{5.2}$$

In 1965 Marvin Bloom and the author showed (see Reference 3 in **[3]**) that

$$K_1 \leqslant \frac{5-2\sqrt{2}}{24}\pi = (.09048)\pi. \tag{5.3}$$

The author even conjectured that in (5.3) we have the equality sign. This conjecture was beautifully demolished by F. Cunningham, who proved in [**3**] the most remarkable

THEOREM 4 (F. Cunningham). *Again*

$$K_1 = 0. \tag{5.4}$$

There are even simply connected polygons, within the unit circle $x^2+y^2 \leqslant 1$, having arbitrarily small area and which are K-sets.

Because the right side of (5.3) is the limit of the areas of certain *star-shaped* K-sets, it follows that

$$K_s \leqslant \frac{5-2\sqrt{2}}{24}. \tag{5.5}$$

This time it is impossible for K_s to slip down to zero again, because of the following remarkable

THEOREM 5 (F. Cunningham).

$$K_s \geqslant \frac{\pi}{108}. \tag{5.6}$$

Combining (5.5) and (5.6) we now know that

$$\frac{\pi}{108} \leqslant K_s \leqslant \frac{5-2\sqrt{2}}{24}\pi. \tag{5.7}$$

Let me now state a few conjectures.

CONJECTURE 1. In Perron's Theorem 2 of §4 we have the equality sign. This means that

$$B_{2^m} = \frac{2}{m+2}. \tag{5.8}$$

CONJECTURE 2. The sequence (B_n) of the Besicovitch constants B_n, defined by (3.3), is strictly decreasing.

CONJECTURE 3. We have

$$\lim_{n\to\infty} B_n \log n = 2\log 2.$$

It is easy to show that $\overline{\lim}\, B_n \log n \leqslant 2\log 2$ (see [**6**]).

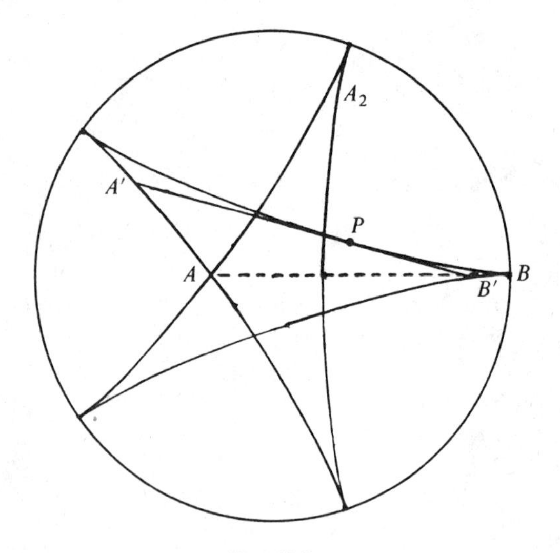

FIG. 13.9

CONJECTURE 4. In (5.7) K_s is equal to the upper bound, so that

$$K_s = \frac{5-2\sqrt{2}}{24}\pi. \tag{5.9}$$

This, if true, is difficult to establish. In a joint paper with F. Cunningham in 1964, listed in [**3**] as the third reference, the author considered certain star-shaped domains A_n having $2n+1$ cusps, composed of $2n+1$ circular arcs which are orthogonal to the circle in which the entire figure is inscribed. Figure 13.9 shows the domain A_2 having 5 cusps. The domain A_n is so scaled that $AB=1$. It was shown there that

$$\lim_{n\to\infty} \text{Area } A_n = \frac{5-2\sqrt{2}}{24}\pi.$$

An attempt at proving Conjecture 4 would require a domain C_n, having the shape and all the symmetries of A_n, with the additional and crucial property that, if P is an arbitrary point of the curve, then

$$A'B' = \text{constant} = 1.$$

The domain C_1 is identical with Kakeya's hypocycloid H.

Do these domains C_n exist if $n>1$?

Problems

1. Establish the equation (3.4), hence $B_2 = 2/3$, by showing that in Figure 13.3 we have the largest possible overlap between the triangles t_1^* and t_2^*.

2. Recall the definition (3.3) of the Besicovitch constant B_n in the first paragraph of §3. If the area of the triangle $T = ABC$ is not necessarily 1, then (3.3) is to be replaced by

$$\min \text{Area} \bigcup_{1}^{n} t_i^* = B_n \cdot \text{Area } T.$$

Use this to show that if $n = n_1 + n_2$, then

$$B_n \leqslant \frac{n_1}{n} B_{n_1} + \frac{n_2}{n} B_{n_2}.$$

Generalize this as follows.

$$\text{If} \quad n = \sum_{1}^{k} n_i, \quad \text{then} \quad B_n \leqslant \frac{1}{n} \sum_{1}^{k} n_i B_{n_i}. \tag{1}$$

Hint: Overlap maximally the triangles $t_1^*, t_2^*, \ldots, t_{n_1}^*$, and also $t_{n_1+1}^*, \ldots, t_{n_1+n_2}^*$, etc.

3. A special shape of the set $P_n = \cup_{i=1}^{n} t_i^*$ which we call "the pineapple" is shown in Figure 13.10 for $n = 5$. Again $T = ABC$ is a 45°-90°-45° triangle of height 1 and area 1. We divide T into n triangles of equal areas,

$$T = \bigcup_{1}^{n} t_i, \tag{2}$$

and mark the intersections $p_0, p_1, \ldots, p_n$ of the t_i with the line $y = 1/3$. Now slide horizontally the t_i together so that all segments $p_i p_{i+1}$ $(i = 0, 1, \ldots, n-1)$ coincide with the segment RS, obtaining on the left in Figure 13.10 the set P.

Show that

$$\text{Area } P_n = \frac{n+2}{3n}. \tag{3}$$

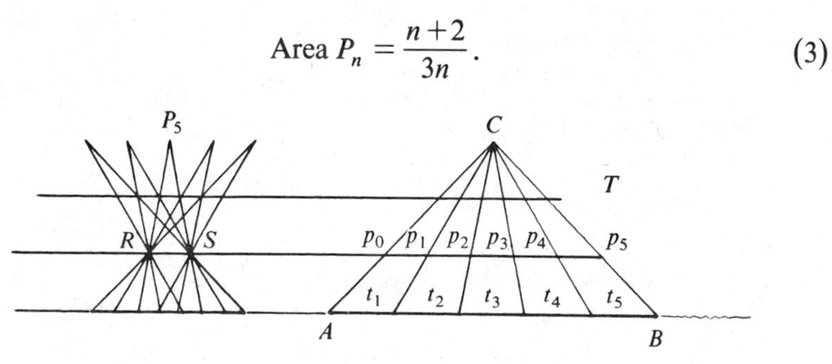

FIG. 13.10

This result shows that Area $P_n \to 1/3$ as $n \to \infty$, so that the pineapple P_n cannot replace T_n in proving Theorem 1.

4. Let t_i^* be the triangle obtained from t_i, of (2), by translating t_i horizontally by an arbitrary amount x_i $(i=1,\ldots,n)$, and let

$$f(x_1, x_2, \ldots, x_n) = \text{Area} \bigcup_1^n t_i^*. \tag{4}$$

Show

(i) That the area (3) of P_n is a *stationary value* of the function f in the sense that for P_n the equations

$$\frac{\partial f}{\partial x_i} = 0 \qquad (i=1,\ldots,n) \tag{5}$$

are satisfied.

(ii) That the equations (5) also hold for the Perron tree T_n of height 1. Actually the height of the tree does not matter, and we may use the trees of Figure 13.7 (see [**4**, p. 363]).

Hint: Look at the set P_5 of Figure 13.10 and in particular to those portions of the lateral sides of t_i^* which are on the boundary of the set P_5. We obtain $\partial f/\partial x_i$ by displacing t_i^* by an infinitesimal amount. Do the same for T_1, T_2, T_3 of Figure 13.7. Can you use induction in n to show that T_n leads to a stationary value of (4)? The example of P_n shows that the function (4) has more than one stationary value.

References

[**1**] A. S. Besicovitch, *On Kakeya's problem and a similar one*, Math. Z., 27 (1928) 312–320.

[**2**] ______, *The Kakeya problem*, Amer. Math. Monthly, 70 (1963) 697–706.

[**3**] F. Cunningham, *The Kakeya problem for simply connected and for star-shaped sets*, Amer. Math. Monthly, 78 (1971) 114–129.

[**4**] O. Perron, *Über einen Satz von Besicovitch*, Math. Z., 28 (1928) 383–386.

[**5**] I. J. Schoenberg, *On the Besicovitch-Perron Solution of the Kakeya Problem*, Studies in Math. Analysis and related topics, Essays in honor of George Pólya, Stanford Univ. Press, Stanford, Calif., 1962, pp. 359–363.

[**6**] ______, *On certain minima related to the Besicovitch-Kakeya problem*, Mathematica (Cluj), 4 (1962) 145–148.

For further developments and references see

[**7**] F. Cunningham, *Three Kakeya problems*, Amer. Math. Monthly, 81 (1974) 582–592.

CHAPTER **14**

ON TWO THEOREMS OF PONCELET AND STEINER

1. Introduction. These are two theorems from the golden age of geometry which brought such advances as projective geometry and the noneuclidean geometries. Here are the theorems.

THEOREM 1 (Poncelet). *Let C and C' be two circles in the plane, C' inside C. We start from a point P on C, and draw from P a tangent to C' in the counterclockwise direction and intersecting C at P_1. From P_1 we draw the tangent to C' intersecting C at P_2, and we continue this construction n times, the last tangent to C' being $P_{n-1}P_n$. If it so happens that*

$$P_n = P \qquad (n \geqslant 3) \tag{1.1}$$

after one revolution around the circle C, then this will also happen no matter where the initial point P has been chosen on C.

THEOREM 2 (Steiner). *Again we have two circles, C and C', with C' inside C. We draw a circle Γ tangent to both C and C'. Next we draw the circle Γ_1 tangent to C, C', and also to Γ. We continue this construction n times, the last circle Γ_n being tangent to C, C', and also to Γ_{n-1}. If it so happens that*

$$\Gamma_n = \Gamma \qquad (n \geqslant 3), \tag{1.2}$$

after one revolution around the circle C', then this will also happen no matter what initial circle Γ has been chosen.

To Theorem 1: If (1.1) holds, then we obtain a closed convex polygon $(P) = PP_1 \ldots P_{n-1}$, which is inscribed in C and circumscribed to C'. Theorem 1 tells us that we can sort of "turn" (P) within the ring (C, C') to give P an arbitrary position on C.

Notice that it is not easy to find circles C and C' admitting a closed polygon (P) both inscribed and circumscribed to C and C'. However, in the simplest case, when $n = 3$, this is easy: We start from a triangle $T = PP_1P_2$, and choose for C its circumcircle, and for C' its incircle. Here it is clear that

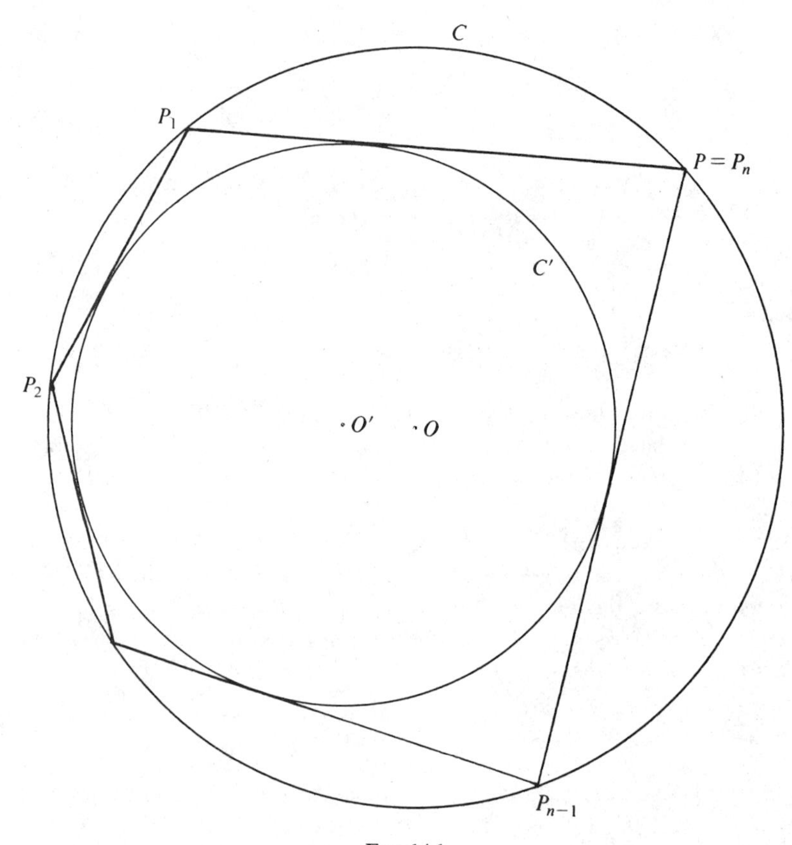

FIG. 14.1

$P_3 = P$, and we conclude from Theorem 1 that we can "sort of turn" the triangle within the fixed ring (C, C').

A common feature of both theorems is that they both become obvious if *the circles C and C' are concentric*. Indeed, the closed polygon $(P) = PP_1 \ldots P_{n-1}$ is then a regular n-gon which can evidently be turned with the circular ring (C, C'). Also, if (1.2) holds, then the chain of circle $(\Gamma) = \Gamma\Gamma_1 \ldots \Gamma_{n-1}$ looks like a tightly fitting chain of roller-bearings which can be turned around. These remarks raise the question:

May our theorems be established by reducing them to the case of concentric circles C and C'? (1.3)

We shall see in §2 that this can be done for Theorem 2 but not for Theorem 1.

In §3 we establish Poncelet's Theorem 1 following J. Bertrand [**1**, pp. 575–577] who attributes the proof to Jacobi. However, examining Jacobi's

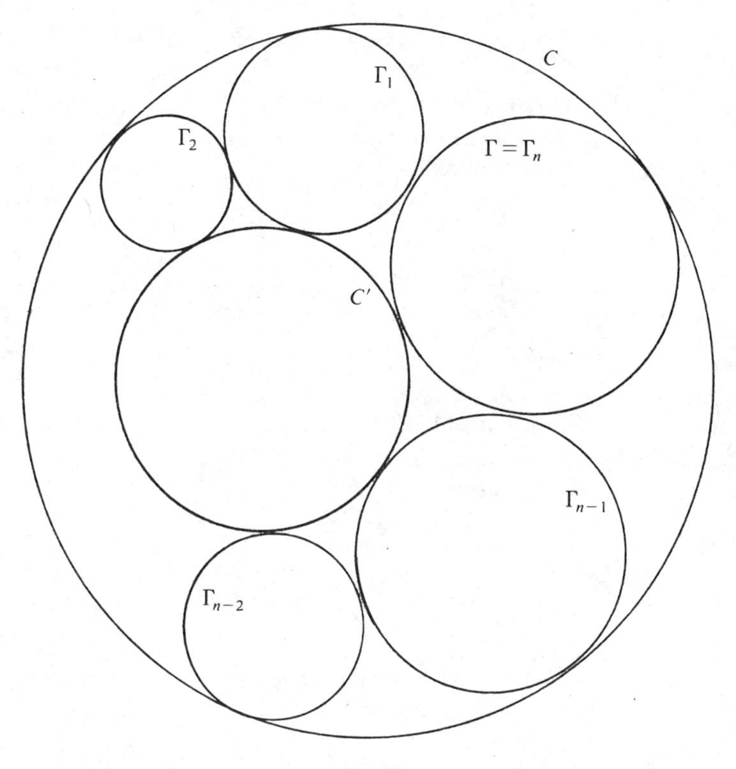

FIG. 14.2

memoir in the first volume of his Collected Works, we realize that the proof should be attributed more to Bertrand than to Jacobi. It is a remarkable application of elementary calculus to geometry.

2. The Answers to the Question (1.3). In trying to answer this question, we would like to transform Figure 14.1 and also Figure 14.2 in such a way that the circles C and C' become concentric. What sort of transformations should be used for this purpose? This will depend on which of the two theorems we wish to establish.

A. *A proof of Theorem* 2. Since Figure 14.2 is composed of circles, we must use such point-to-point transformations that transform circles into circles. Such transformations are the linear fractional, or Moebius, transformations of the form

$$w = \frac{Az + B}{Cz + D} \qquad (A, B, C, D \text{ are complex constants}). \qquad (2.1)$$

These are discussed in any introduction to functions of a complex variable $z = x + iy$.

Let D and D' denote the circular discs bounded by C and C', respectively. We assume that C and C' are not concentric and, without loss of generality, we assume that D is the unit circle

$$D: |z| \leqslant 1. \tag{2.2}$$

Moreover, by an appropriate rotation $w = e^{i\alpha}z$ we may assume the center of D' to be on the segment $0 < x < 1$. Finally, let the intersection

$$[\alpha, \beta] = D' \cap \{y = 0\} \qquad (\alpha + \beta > 0) \tag{2.3}$$

be a diameter of D' (see Figure 14.3).

Is there a transformation (2.1) *mapping D onto itself and such that D' is mapped into the* (concentric) *circle*

$$|z| \leqslant r \qquad (0 < r < 1)\,? \tag{2.4}$$

The affirmative answer is obtained as follows. We first observe that the mapping

$$w = \frac{z + a}{az + 1} \qquad (-1 < a < 1) \tag{2.5}$$

maps D onto itself, such that $z = -1, 0, 1$ are mapped into $w = -1, a, 1$, respectively. We wish to show that we can find a such that α and β are mapped into the points $-r$ and r, respectively. Using (2.5), this amounts to the equation

$$\frac{\alpha + a}{a\alpha + 1} = -\frac{\beta + a}{a\beta + 1},$$

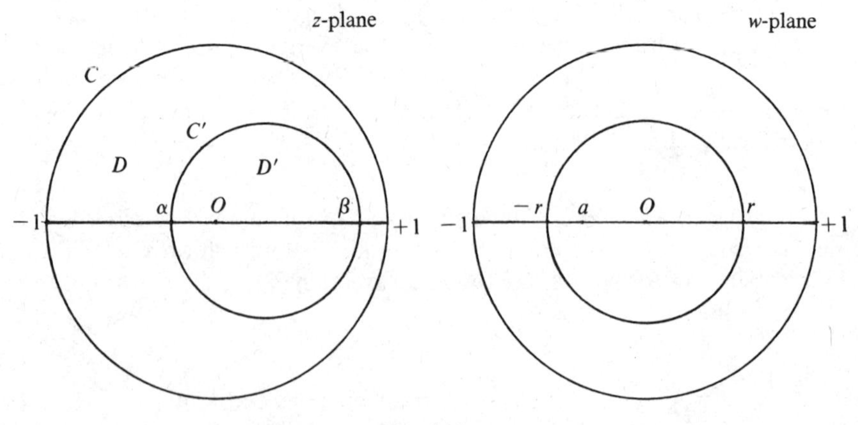

Fig. 14.3

which is found to be equivalent to the quadratic equation in a;

$$a^2+2a\frac{1+\alpha\beta}{\alpha+\beta}+1=0. \tag{2.6}$$

This equation has real roots, because its discriminant

$$\left(\frac{1+\alpha\beta}{\alpha+\beta}\right)^2-1=\frac{1}{(\alpha+\beta)^2}(1-\alpha^2-\beta^2+\alpha^2\beta^2)$$

$$=\frac{1}{(\alpha+\beta)^2}(1-\alpha^2)(1-\beta^2)>0$$

is seen to be positive. Since the product of the two roots of (2.6) is 1, we conclude that there is a unique a, such that

$$-1<a<0, \tag{2.7}$$

the last inequality (2.7) following from the fact that the two roots of (2.6) are negative. With this value of a we see that (2.5) maps D' into the disk (2.4), and Theorem 2 is established. Concerning Theorem 2, see also **[3]**.

B. The present discussion may be omitted because its result is a negative one: *It aims to show that concerning Theorem* 1 *the answer to question* (1.3) *is negative*. In Poncelet's case we need transformations of the disk D, of (2.2), which map straight lines into straight lines. These are (real) transformations of the form

$$x'=\frac{ax+by+c}{a''x+b''y+c''},\quad y'=\frac{a'x+b'y+c'}{a''x+b''y+c''}. \tag{2.8}$$

The analogue of (2.5) is now the transformation

$$P_a: x'=\frac{x+a}{ax+1},\quad y'=\frac{y}{ax+1}\sqrt{1-a^2}\qquad(-1<a<1), \tag{2.9}$$

which maps the disk $D: x^2+y^2\leqslant 1$, onto itself, such that the points

$$(-1,0),(0,0),(1,0) \text{ are mapped into } (-1,0),(a,0),(1,0),$$

respectively. The inverse transformation is found to be

$$P_a^{-1}:\quad x=\frac{x'-a}{1-ax'},\quad y=\frac{y'}{1-ax'}\sqrt{1-a^2}. \tag{2.10}$$

The image of the circle $x^2+y^2=r^2$ $(0<r<1)$, is easily found to be the ellipse

$$\frac{(x'-\xi)^2}{A^2}+\frac{y'^2}{B^2}=1, \tag{2.11}$$

where

$$\xi=\frac{1-r^2}{1-a^2r^2}a,\quad A=\frac{(1-a^2)r}{1-a^2r^2},\quad B=r\sqrt{\frac{1-a^2}{1-a^2r^2}} \tag{2.12}$$

(see Figure 14.4). I claim: *If* $0<r<1$ *and* $0<a<1$, *then* (2.11) *is an ellipse and is never a circle.*

Indeed, let us show that

$$A<B. \tag{2.13}$$

From (2.12) we find (2.13) to amount to

$$\frac{r^2(1-a^2)^2}{(1-a^2r^2)^2}<r^2\frac{1-a^2}{1-a^2r^2},$$

and, finally, to $1-a^2<1-a^2r^2$, which is evident.

We have just shown that the two nonconcentric circles C and C' cannot be changed into concentric ones by a projective transformation. It follows that Theorem 1 requires a different approach to be given in §3.

C. *The two euclidean models of noneuclidean geometry.* It seems worthwhile to mention the connection of our discussion with noneuclidean geometry.

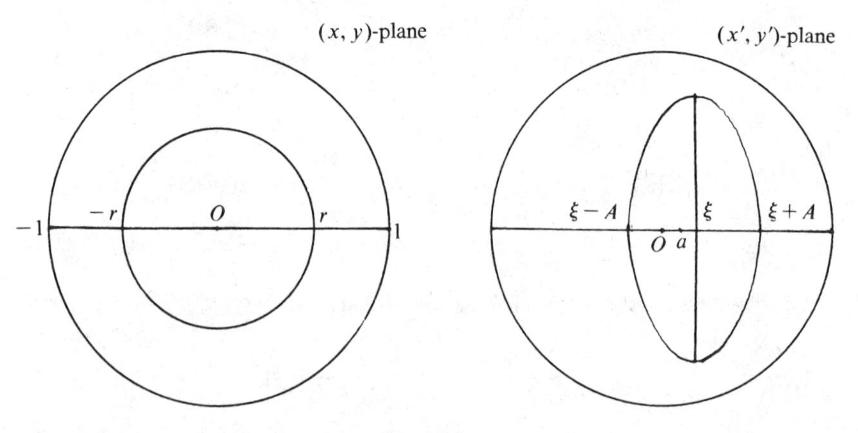

FIG. 14.4

The projective transformations of the disk D onto itself form the group of noneuclidean rigid motions in F. Klein's euclidean model of noneuclidean geometry. In contrast, the Moebius transformations of D onto itself form the noneuclidean rigid motions in H. Poincaré's euclidean model of noneuclidean geometry. For a discussion of these models see [**2**, Chapter 4, §9].

3. The Bertrand-Jacobi Proof of Poncelet's Theorem 1. We are following the proof as given by Bertrand [**1**, pp. 575–577]. Figure 14.5 shows the circles C and C', having centers O and O', and the first tangent PP_1, touching C' at M. Let

$$OA = R, \quad O'M = r, \quad O'O = a,$$
$$\angle AOP = 2\phi, \quad \angle AOP_1 = 2\phi_1. \tag{3.1}$$

The fundamental observation is that *the rate of change of* ϕ_1 *as a function of* ϕ *is expressed as*

$$\frac{d\phi_1}{d\phi} = \frac{MP_1}{MP}. \tag{3.2}$$

This we derive as follows. Let $P'M'P_1'$ be a neighboring tangent to C', intersecting PMP_1 in N. From the similitude of the triangles PNP' and $P_1'NP_1$ we obtain that

$$\frac{P_1P_1'}{PP'} = \frac{NP_1}{NP'}. \tag{3.3}$$

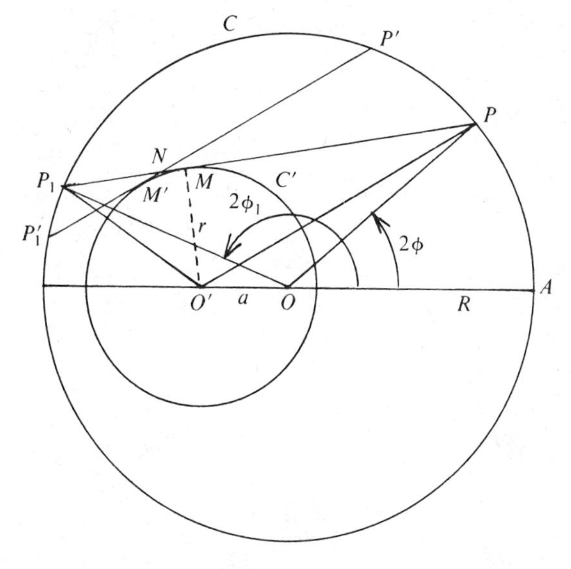

FIG. 14.5

From $\widehat{PP'} = 2R\Delta\phi$, $\widehat{P_1P_1'} = 2R\Delta\phi_1$, we conclude that, as $\Delta\phi \to 0$, for the limit of the left-hand side of (3.3) we have

$$\lim \frac{P_1P_1'}{PP'} = \lim \frac{\widehat{P_1P_1'}}{\widehat{PP'}} = \lim \frac{\Delta\phi_1}{\Delta\phi} = \frac{d\phi_1}{d\phi},$$

while for the right-hand side of (3.3)

$$\lim \frac{NP_1}{NP'} = \frac{MP_1}{MP},$$

and (3.2) is established.

Let us evaluate the lengths MP and MP_1. Clearly

$$\begin{aligned}(O'P)^2 &= R^2 + a^2 + 2aR\cos 2\phi \\ &= (R^2+a^2)(\cos^2\phi + \sin^2\phi) + 2aR(\cos^2\phi - \sin^2\phi) \\ &= (R+a)^2\cos^2\phi + (R-a)^2\sin^2\phi \\ &= (R+a)^2(1-\sin^2\phi) + (R-a)^2\sin^2\phi\end{aligned}$$

and so

$$(O'P)^2 = (R+a)^2 - 4aR\sin^2\phi.$$

Now

$$MP = \sqrt{(O'P)^2 - r^2} = \sqrt{(R+a)^2 - r^2 - 4aR\sin^2\phi}.$$

Setting

$$k^2 = \frac{4aR}{\sqrt{(R+a)^2 - r^2}} \tag{3.4}$$

we finally obtain that

$$MP = \sqrt{(R+a)^2 - r^2} \cdot \sqrt{1 - k^2\sin^2\phi}. \tag{3.5}$$

By similar calculations we conclude that

$$MP_1 = \sqrt{(R+a)^2 - r^2} \cdot \sqrt{1 - k^2\sin^2\phi_1}. \tag{3.6}$$

From (3.2), (3.5), and (3.6), we obtain the differential equation

$$\frac{d\phi_1}{d\phi} = \frac{\sqrt{1-k^2\sin^2\phi_1}}{\sqrt{1-k^2\sin^2\phi}},$$

which we rewrite as

$$\frac{d\phi_1}{\sqrt{1-k^2\sin^2\phi_1}} = \frac{d\phi}{\sqrt{1-k^2\sin^2\phi}}. \tag{3.7}$$

We now consider the integral

$$J(\phi,\phi_1) = \int_{\phi}^{\phi_1} \frac{d\theta}{\sqrt{1-k^2\sin^2\theta}}. \tag{3.8}$$

By the rule of differentiation of definite integrals with variable limits of integration, we conclude that its differential is

$$dJ(\phi,\phi_1) = \frac{d\phi_1}{\sqrt{1-k^2\sin^2\phi_1}} - \frac{d\phi}{\sqrt{1-k^2\sin^2\phi}}$$

and therefore $dJ(\phi,\phi_1)=0$ in view of (3.7). We conclude that

$$J(\phi,\phi_1) = \omega \textit{ is a constant which is independent of } \phi. \tag{3.9}$$

This will readily yield a proof of Theorem 1. Indeed, if we now perform the constructions of Theorem 1 obtaining the angles

$$\angle AOP_2 = 2\phi_2,\ \angle AOP_3 = 2\phi_3,\ \ldots,\ \angle AOP_n = 2\phi_n, \tag{3.10}$$

we obtain, by (3.8) and (3.9), that

$$\int_{\phi}^{\phi_n} \frac{d\theta}{\sqrt{1-k^2\sin^2\theta}} = J(\phi,\phi_1) + J(\phi_1,\phi_2) + \cdots + J(\phi_{n-1},\phi_n).$$

Since (3.9) implies that

$$J(\phi_\nu,\phi_{\nu+1}) = \omega \quad \text{for } \nu = 0,\ldots,n-1, \quad \text{where } \phi_0 = \phi,$$

we conclude that

$$\int_{\phi}^{\phi_n} \frac{d\theta}{\sqrt{1-k^2\sin^2\theta}} = n\omega. \tag{3.11}$$

However, the assumption of Theorem 1 is that $P_n = P$ after one revolution along C, and this means by (3.10) that $2\phi_n = 2\phi + 2\pi$, and therefore

$$\phi_n = \phi + \pi. \tag{3.12}$$

The assumption of Theorem 1 therefore implies the equation

$$\int_{\phi}^{\phi+\pi} \frac{d\theta}{\sqrt{1-k^2\sin^2\theta}} = n\omega. \tag{3.13}$$

This, then, is an equation implied by our assumption that $P = P_n$. The converse is also true because the integrand of (3.13) is positive, and therefore (3.13) means, by (3.10), that

$$\phi_n - \phi = \pi.$$

The question now is this: *Does the equation* (3.13) *remain valid if we vary* ϕ? Evidently so, because the integrand of (3.13) is a periodic function of period π, and so the integral over a period, from ϕ to $\phi + \pi$, does not depend on the value of ϕ. This completes our proof of Theorem 1.

If we should have that $P_n = P$, not after one revolution, but after k revolutions around the circle C, then (3.13) is to be replaced by

$$\int_{\phi}^{\phi+k\pi} \frac{d\theta}{\sqrt{1-k^2\sin^2\theta}} = n\omega$$

and again the equation remains valid if we vary ϕ.

The beauty of the proof is that, while it uses the elliptic integral

$$\int \frac{d\theta}{\sqrt{1-k^2\sin^2\theta}},$$

it uses only the simplest concepts of integral calculus.

4. A Generalization of Theorem 1′. Let us project Figure 14.1 from a point V (like a source of light-rays) onto another plane π, so that the circles C and C' are projected into ellipses Γ and Γ', with Γ' in the interior of Γ. It seems then obvious that Theorem 1 will remain valid in the plane π, with the ellipses Γ and Γ' playing the role of the circles C and C'. This remark suggests the following general

THEOREM 1′ (Poncelet). *Theorem* 1 *remains valid if we replace the circles C and C' by any two ellipses Γ and Γ' such that Γ' is in the interior of Γ.*

For a proof see [**4**].

Problems

1. Let ABC be a triangle with circumcircle C of radius R, and incircle C' of radius r. If $OO'=a$ denotes the distance between their centers, show that

$$a^2 = R(2r-R). \tag{1}$$

This theorem is due to L. Euler.

Hint: We use Poncelet's theorem for $n=3$ and can therefore so "turn" ABC, keeping C and C' fixed, that the vertex A is on the line OO' joining the centers. This reduces the problem to the case when ABC is an isosceles triangle (Figure 14.6(a)). Let $\angle BAC=\theta$, and evaluate AB in terms of R and θ. Derive two equations between R, r, a, and θ from the two triangles $BO'E$ and BDA. Obtain Euler's Theorem (1) by eliminating θ between these two equations. From (1) we conclude that $2r>R$, unless $2r=R$; hence $a=0$ and the triangle ABC is equilateral.

2. Let $Q=ABCD$ be a convex quadrilateral which is inscribed in a circle C of radius R and circumscribed to a circle C' of radius r. If $OO'=a$ denotes the distance between the centers, show that we have the equation

$$(R^2-a^2)^2 = 2r^2(R^2+a^2). \tag{2}$$

Hint: Again we use Poncelet's theorem and "turn" Q, keeping the circles C and C' fixed, so that AD becomes perpendicular to the line OO' joining the centers. Then BC *must* become parallel to AB, as the quadrilateral would otherwise not close, as it must. This reduces the problem to the case

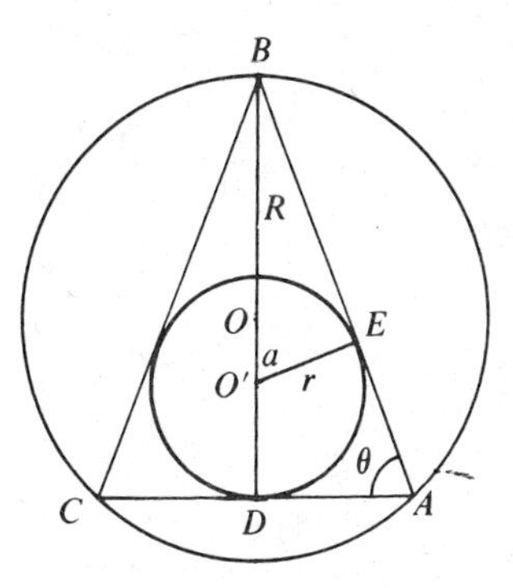

(a)

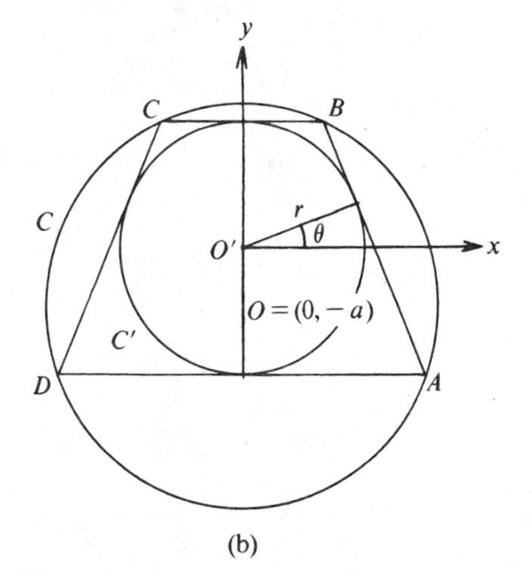

(b)

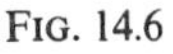
FIG. 14.6

of an isosceles trapezoid $ABCD$. See Figure 14.6(b). Choose a coordinate system with O' as origin. Let the line AB have the equation $x\cos\theta + y\sin\theta = r$, and determine the points A and B. Let $O = (0, -a)$ be such that $R^2 = (OA)^2$ and $R^2 = (OB)^2$. Eliminate θ between these two equations to obtain (2).

3. Let S be the surface of a sphere with center O. Let C be a small circle on S, bounding the small spherical cap $\bar{C}$. Let C' be a small circle in the interior of $\bar{C}$. Show that Theorem 1 remains valid for the two spherical circles C and C', where the role of the line segments $PP_1, P_1P_2, \ldots$ of Theorem 1 is taken over by arcs of great circles.

Hint: Let $K = \text{Cone}\,(O, C)$ be the circular cone containing C with vertex O, and $K' = \text{Cone}\,(O, C')$. Let π be a plane intersecting K and K', and apply Theorem 1′ to the two ellipses $\Gamma = \pi \cap K$, $\Gamma' = \pi \cap K'$.

References

[1] J. Bertrand, *Traité de calcul différentiel et de calcul intégral*, vol. 1, Gauthier-Villars, Paris 1870.

[2] R. Courant and H. Robbins, *What Is Mathematics*?, Oxford University Press, London-New York-Toronto, 1941.

[3] D. T. Piele, M. W. Firebaugh, and R. Manulik, *Applications of conformal mapping to potential theory through computer graphics*. Amer. Math. Monthly, 84 (1977) 677–692.

[4] I. J. Schoenberg, *On Jacobi's proof of a theorem of Poncelet*, to appear in a volume dedicated to Paul Turán by the Hungarian Academy of Sciences.

CHAPTER **15**

ON THE MOTIONS OF A BILLIARD BALL I: INTRODUCING THE KÖNIG-SZÜCS POLYGONS AND POLYHEDRA

1. Introduction. The present MTE and also the next three deal with a special kind of polygons and polyhedra introduced by D. König and A. Szücs in a paper [**2**] with the romantic-sounding title "The motion of a point abandoned in the interior of a cube." Accordingly, we shall call these geometric objects *König-Szücs polygons* and *König-Szücs polyhedra*, abbreviating the names of the authors to K-S. They arise from the following dynamical problem. A perfectly elastic material particle x (called a billiard ball, abbreviated to b.b.) moves inside a fixed unit cube γ_3. As we assume that x moves within γ_3 in a vanishing field of force, we see that the motion of x is rectilinear and uniform, while being reflected in the usual way on striking any of the six facets of the cube γ_3. These peculiar conditions have the effect that instead of a problem on differential equations to which dynamical problems usually lead, we have before us arithmetical problems.

Nearly one hundred years ago L. Kronecker discovered results on Diophantine Approximations, special cases of which we derive in our next MTE, following the geometric approach of F. Lettenmeyer [**3**]. König and Szücs found that Kronecker's theorems led to similar properties of b.b. motions. This transition may roughly be compared to the passage from the first to the second diagram of Figure 15.1: The second and fourth slanting segments of the first diagram are flipped over to produce the *continuous* M-shaped figure of the second. In terms of the intuitive and continuous billiard ball motions, the results are easier to describe and to visualize. For a discussion of Kronecker's theorems and their application to b.b. motions in γ_2 see also Chapter 23 of [**1**].

A K-S polygon is the geometric path of a billiard ball. We shall also study such motions in two dimensions, within a square "table" γ_2. Examples of such polygons within γ_2 are shown in Figure 15.2(a),(b),(c). Some reader may have seen a design like (a) on a window of a suburban home, perhaps in California. In our next MTE we will see that the square path of (c) enjoys a certain extremum property among all planar K-S polygons.

What are the characteristic properties of the K-S polygons in γ_2? Evidently, they are polygons having all their vertices on the frame of γ_2, and

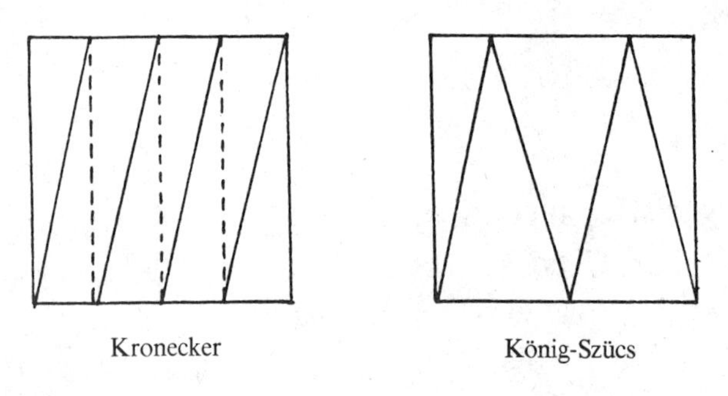

FIG. 15.1

two consecutive sides form equal angles with the side of γ_2 that they strike. The examples of Figure 15.2 are deceivingly simple, as K-S polygons usually have infinitely many sides.

Few readers, if any, have seen K-S polygons within γ_3. In Figure 15.3(a) and (b) we show strikingly simple examples. (See [**4**, Figure 2, on p. 278].)

1. The first, shown in (a), is a closed hexagon 123456 whose vertices are on the indicated diagonals of the facets of γ_3, and divide these diagonals in the ratio of 1:2.

2. The second, shown in (b), is a closed octagon 12345678; the vertices 2,4,6,8 are the centers of the lateral facets of γ_3, while 3, and also 7, divide a diagonal of the lowest facet in the ratio of 1:3. Likewise 1 and 5 divide a diagonal of the top facet in the same ratio.

The contents of this MTE are as follows. In §2, we introduce an appropriate auxiliary function, the so-called *linear Euler spline*, that will furnish neatly and automatically the b.b. motion defined by a given initial motion. The *Corner Reflectors*, of importance in astronomical radar applications, are described in §3. Theorem 1, of §4, characterizes the finite K-S polygons. In §5 we mention, without proofs, the striking results of König

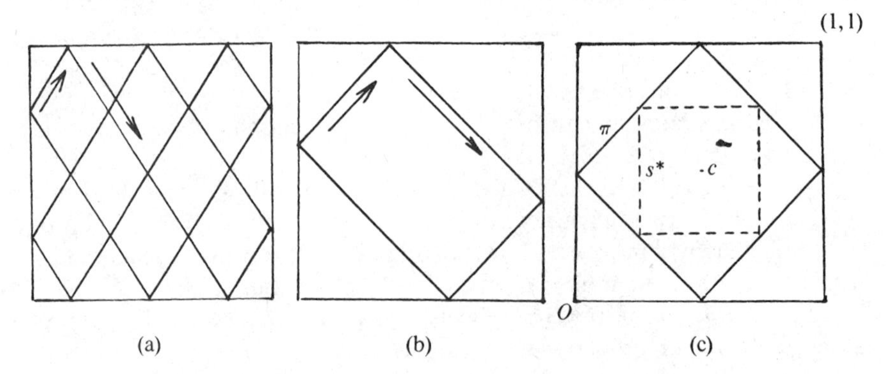

FIG. 15.2

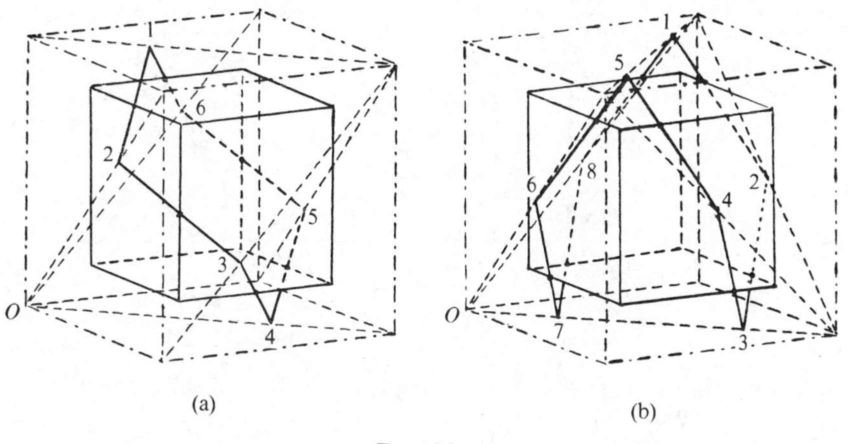

FIG. 15.3

and Szücs on the set-theoretic closures of infinite K-S polygons, and use them to motivate the introduction of the K-S *polyhedra*. These are obtained by replacing the billiard ball by a flat pencil of light-rays issuing from a point-source, which is reflected by the facets of γ_3, thought of as mirrors. Finally, in §6 the closed and finite K-S polyhedra are characterized in Theorem 2.

2. A Parametric Representation of K-S Polygons. In a Cartesian coordinate system $Ox_1x_2x_3$, let

$$\gamma_3 : 0 \leq x_i \leq 1 \qquad (i=1,2,3) \tag{2.1}$$

denote the unit cube within which b.b. motions are to be studied. Furthermore, let

$$L_1 : x_i = \lambda_i t + a_i \qquad (i=1,2,3) \quad (-\infty < t < \infty) \tag{2.2}$$

be the initial rectilinear and uniform motion of the point $x=(x_1, x_2, x_3)$. This is the point of König and Szücs which is "abandoned" in the interior of γ_3. About its constant velocity vector $\lambda=(\lambda_1, \lambda_2, \lambda_3)$ we naturally assume that

$$|\lambda|^2 = \lambda_1^2 + \lambda_2^2 + \lambda_3^2 > 0. \tag{2.3}$$

We can also adjust in (2.2) the time parameter t such that the initial position $a=(a_1, a_2, a_3)$ is in the interior of γ_3.

In order to obtain the required reflections of x we use the *linear Euler spline* $\langle x \rangle$ defined by the relations

$$\langle x \rangle = |x| \quad \text{if} \quad -1 \leq x \leq 1 \quad \text{and} \quad \langle x+2 \rangle = \langle x \rangle \quad \text{for all real } x. \tag{2.4}$$

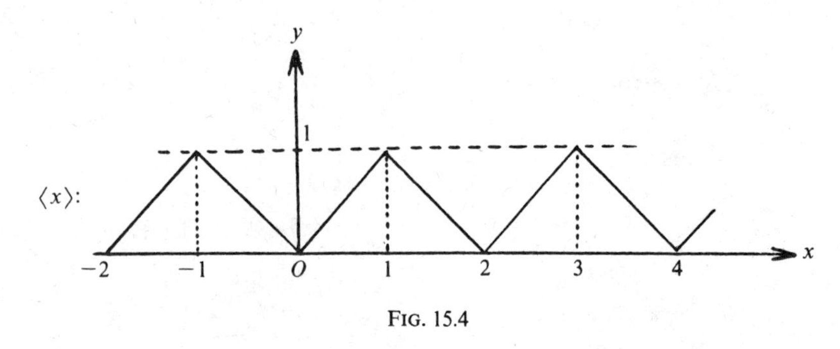

FIG. 15.4

This is an even function of period 2 having a piecewise linear graph shown in Figure 15.4. We may now describe

A DEFINITION OF BILLIARD BALL MOTIONS. *The b.b. motion within* γ_3 *for the initial motion* (2.2) *is defined parametrically by the equations*

$$\Pi^1: \begin{aligned} x_1 &= \langle\lambda_1 t + a_1\rangle, \\ x_2 &= \langle\lambda_2 t + a_2\rangle, \\ x_3 &= \langle\lambda_3 t + a_3\rangle, \qquad (-\infty < t < \infty). \end{aligned} \tag{2.5}$$

REMARKS. 1. Because $0 \leqslant \langle x\rangle \leqslant 1$ for all real x, it is clear that the equations (2.5) define a motion within the cube γ_3. Moreover, from $0 < a_i < 1$ $(i = 1,2,3)$ and the graph of Figure 15.4 we see that the motions (2.2) and (2.5) are identical as long as $|t|$ is sufficiently small. This seems neat enough, but there still remains the question: *Does the motion* (2.5) *perform the required reflections in the facets of* γ_3?

2. To justify an affirmative answer we observe the following.

(i) *If some of the* λ_i *vanish, then we have a lower-dimensional motion.* For instance, if $\lambda_1 \neq 0$, while $\lambda_2 = \lambda_3 = 0$, then (2.5) describes a to-and-fro motion

$$x_1 = \langle\lambda_1 t + a_1\rangle, \quad x_2 = a_2, \quad x_3 = a_3$$

on the segment $0 \leqslant x_1 \leqslant 1, x_2 = a_2, x_3 = a_3$. If $\lambda_1 \neq 0$, $\lambda_2 \neq 0$, $\lambda_3 = 0$, then (2.5) become

$$x_1 = \langle\lambda_1 t + a_1\rangle, \quad x_2 = \langle\lambda_2 t + a_2\rangle, \quad x_3 = a_3,$$

and this is a b.b. motion in the unit square $\gamma_3 \cap \{x_3 = a_3\}$.

(ii) Let us now assume that no λ_i vanishes, for instance that

$$\lambda_1 > 0, \quad \lambda_2 > 0, \quad \lambda_3 > 0.$$

As t increases, the increasing linear function $\lambda_i t + a_i$ will successively and periodically assume *even integer values* and *odd integer values*, and the graphs of $\lambda_i t + a_i$ and $\langle \lambda_i t + a_i \rangle$ are shown in Figure 15.5, showing that the straight graph of $\lambda_i t + a_i$ is changed into the zig-zag graph of $\langle \lambda_i t + a_i \rangle$. It follows that if *just one* of the right-hand sides of (2.5) vanishes for $t = t_0$, say $\langle \lambda_3 t_0 + a_3 \rangle = 0$, then

$$\langle \lambda_3 t + a_3 \rangle = \begin{cases} -\lambda_3(t - t_0) & \text{if} \quad t < t_0, \\ \lambda_3(t - t_0) & \text{if} \quad t > t_0, \end{cases}$$

and the effect is a reflection of the b.b. motion in the plane $x_3 = 0$. It also follows that if $\langle \lambda_3 t + a_3 \rangle$ reaches the value 1 for $t = t_0$, while $\langle \lambda_i t_0 + a_i \rangle$ $(i = 1, 2)$ are in the interval $(0, 1)$, then

$$\langle \lambda_3 t + a_3 \rangle = \begin{cases} 1 + \lambda_3(t - t_0) & \text{if} \quad t < t_0, \\ 1 - \lambda_3(t - t_0) & \text{if} \quad t > t_0. \end{cases}$$

The effect is a reflection of the motion in the plane $x_3 = 1$. Our discussion also explains what happens if two, or three, of the right-hand sides of (2.5) reach the values 0 or 1 simultaneously for $t = t_0$, i.e., when *the* b.b. *strikes an*

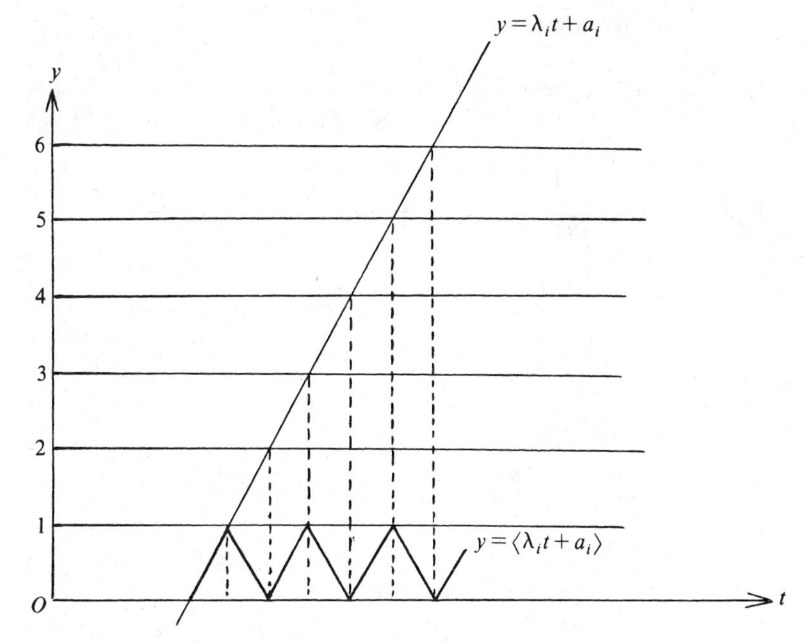

FIG. 15.5

edge of γ_3, *or even a vertex.* The three possibilities are shown in Figure 15.6(a), (b), and (c). In (b) we have a reflection in the edge OA. The incident path, the reflected path, and the edge OA are all in the same plane; moreover, the two paths form equal angles with the perpendicular PN to the edge OA; in (c) we have reflection in the vertex O and the b.b. returns on the same segment on which it arrived at O.

3. *The Corner Reflectors.* An important application of b.b. motions is as follows. Let $Ox_1x_2x_3$ be the positive octant of a rectangular coordinate system, where we assume the boundary quadrants of the octant to be reflecting surfaces. Let

$$x_i = \lambda_i t + a_i \qquad (i=1,2,3;\ t \geqslant 0), \tag{2.6}$$

where

$$\lambda_1 < 0, \quad \lambda_2 < 0, \quad \lambda_3 < 0,$$

be an optical (or radar) signal S coming from within the octant and traveling towards O. Let $S=\overrightarrow{AB}$ strike the plane x_1Ox_2 at B, then x_1Ox_3 at C, and finally x_2Ox_3 at D, returning into the octant along the ray $\overrightarrow{DE}$ (see Figure 15.7). What is the relation between the original direction $\overrightarrow{AB}$ of S and its reflected direction $\overrightarrow{DE}$? The answer is simple and important:

The reflected ray $\overrightarrow{DE}$ *is parallel to and of opposite direction to the incident ray* $\overrightarrow{AB}$. (2.7)

The reason seems obvious: the reflection of the signal (2.6) in the plane $x_j=0$ changes λ_j into $-\lambda_j$. As we reflected the signal in all three planes,

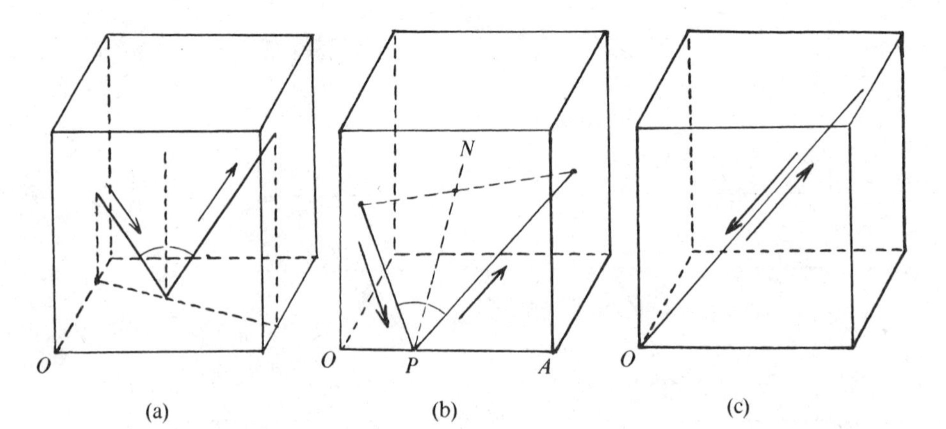

FIG. 15.6

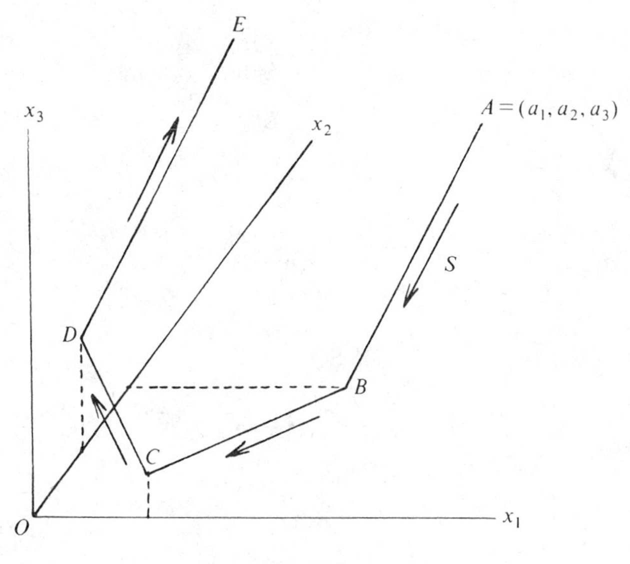

FIG. 15.7

the final returning signal is given by

$$x_i = -\lambda_i t + b_i \qquad (i=1,2,3).$$

Such devices, called *corner reflectors*, were placed on the moon by astronauts and are used to return to earth radar signals sent to our satellite. They allow us to measure very accurately the distance from the terrestrial radar station to the moon's corner reflectors, revealing the size of the tidal motions of the earth's solid crust.

In the special case of Figure 15.6(c), the signal is simply reversed. Another illustration of the property (2.7) can be seen in Figure 15.3(b): the signal $\overrightarrow{56}$ strikes three intersecting facets of the cube at 6, 7, and 8, and emerges as $\overrightarrow{81}$, which is parallel and opposite to $\overrightarrow{56}$.

3. The Finite K-S Polygons. The simplest K-S polygons are the finite ones, and Theorem 1 below states the conditions for their occurrence. However, before we state it, we ask: Why is the hexagon of Figure 15.3(a) a K-S polygon? A verification is easy: From Figure 15.3(a) the coordinates of the vertices 1 and 2 are found to be $(\frac{1}{3}, \frac{1}{3}, 1)$ and $(\frac{2}{3}, 0, \frac{2}{3})$, whence $\overrightarrow{12} = (\frac{1}{3}, -\frac{1}{3} - \frac{1}{3})$. As initial motion along the segment 12 we may therefore take $x_1 = \frac{1}{3} + \frac{1}{3}t$, $x_2 = \frac{1}{3} - \frac{1}{3}t$, $x_3 = 1 - \frac{1}{3}t$. Replacing t by $3t$ we may as well write $x_1 = \frac{1}{3} + t$, $x_2 = \frac{1}{3} - t$, $x_3 = 1 - t$. By (2.5), the equations of our hexagon may be written as

$$x_1 = \langle \tfrac{1}{3} + t \rangle, \quad x_2 = \langle \tfrac{1}{3} - t \rangle, \quad x_3 = \langle 1 - t \rangle. \tag{3.1}$$

Since all these functions have the period 2, the entire polygon is obtained if we restrict t to the interval $0 \leqslant t \leqslant 2$. The right-hand sides of (3.1) assume the values 0 or 1 for precisely the six values

$$t = 0, \tfrac{1}{3}, \tfrac{2}{3}, 1, \tfrac{4}{3}, \tfrac{5}{3}.$$

Substituting these successively in (3.1) we obtain the vertices 1,2,3,4,5,6, in this order. Comparing (3.1) and (2.5) we find that $\lambda_1 = 1, \lambda_2 = -1, \lambda_3 = -1$, and the following theorem will come as no great surprise.

THEOREM 1 (König-Szücs). *The* K-S *polygon*

$$\Pi^1: x_1 = \langle \lambda_1 t + a_1 \rangle, \quad x_2 = \langle \lambda_2 t + a_2 \rangle, \quad x_3 = \langle \lambda_3 t + a_3 \rangle \quad (3.2)$$

$$\left(\sum_1^3 \lambda_i^2 > 0 \right),$$

is a closed and finite polygon if and only if

$$\lambda_1, \lambda_2, \textit{and } \lambda_3 \textit{ are proportional to three rational numbers.} \quad (3.3)$$

(i) *The conditions are sufficient*, for suppose that in terms of integers we have

$$\lambda_1 = \frac{m_1}{q}, \quad \lambda_2 = \frac{m_2}{q}, \quad \lambda_3 = \frac{m_3}{q}, \quad (3.4)$$

with

$$q > 0, \qquad (m_1, m_2, m_3, q) = 1.$$

Each of the three functions (3.2) has the period $2q$, for

$$\langle \lambda_i (t + 2q) + a_i \rangle = \langle \lambda_i t + a_i + 2m_i \rangle = \langle \lambda_i t + a_i \rangle \qquad (i = 1,2,3),$$

and Π^1 is finite and already obtained if we restrict t to $0 \leqslant t \leqslant 2q$.

(ii) *The conditions are necessary*. Indeed, if Π^1 is closed, having the period T, then we must have the three identities

$$\langle \lambda_i t + a_i + \lambda_i T \rangle = \langle \lambda_i t + a_i \rangle \qquad (i = 1,2,3).$$

These imply that for appropriate integers m_i we have $\lambda_i T = 2m_i$ whence

$$\lambda_i = \frac{2}{T} m_i \qquad (i = 1,2,3).$$

Hence $\lambda_1 : \lambda_2 : \lambda_3 = m_1 : m_2 : m_3$, and the proof is complete.

4. The K-S Polyhedra Π^2. We start with a b.b. motion $(-\infty<t<\infty)$

$$\Pi^1: x_1 = \langle \lambda_1 t + a_1 \rangle, \quad x_2 = \langle \lambda_2 t + a_2 \rangle, \quad x_3 = \langle \lambda_3 t + a_3 \rangle, \tag{4.1}$$
$$(\lambda_1 \lambda_2 \lambda_3 \neq 0).$$

In order to motivate the appearance of K-S polyhedra, we assume that Π^1 is *not* a closed and finite polygon. By Theorem 1 this requires that

$$\lambda_1, \lambda_2, \lambda_3 \textit{ are not proportional to three rational numbers}, \tag{4.2}$$

and therefore Π^1 is a polygon having infinitely many sides.

Let (Π^1) denote the polygon Π^1 viewed as the set of points given parametrically by the equations (4.1). Anticipating results to be discussed in our next MTE, we wish to state the following results of König and Szücs. We recall that a point $\zeta \in \gamma_3$ is called a *limit point* of the set (Π^1), provided that every neighborhood of ζ contain points of (Π^1) different from ζ. The union of (Π^1) with the set of all its limit points is called the *closure* of (Π^1) and denoted by $(\overline{\Pi^1})$. König and Szücs have shown that the closure $(\overline{\Pi^1})$ may exhibit only one of the following two totally different aspects:

(i) *The set* $(\overline{\Pi^1})$ *is identical with the entire cube* γ_3; *hence*

$$\left(\overline{\Pi^1}\right) = \gamma_3. \tag{4.3}$$

We also say that (Π^1) is *dense* in γ_3; using a term borrowed from Physics we say that the motion Π^1 is *ergodic* in γ_3.

(ii) *The set* $(\overline{\Pi^1})$ *is identical with a certain finite and closed polyhedron* Π^2; *hence*

$$\left(\overline{\Pi^1}\right) = \Pi^2. \tag{4.4}$$

Of course, Π^2 is a very special kind of polyhedron, called a K-S *polyhedron*, which is defined as follows. We replace in our discussion of §2 the moving billiard ball (2.2) by a flat pencil of light-rays starting from a point $a=(a_1, a_2, a_3)$ of a plane L_2 and moving within this plane. If the parametric equations of this plane are

$$\begin{aligned} x_1 &= \lambda_1^1 t_1 + \lambda_1^2 t_2 + a_1, \\ L_2: x_2 &= \lambda_2^1 t_1 + \lambda_2^2 t_2 + a_2, \\ x_3 &= \lambda_3^1 t_1 + \lambda_3^2 t_2 + a_3, \end{aligned} \tag{4.5}$$

we set

$$t_1 = t\cos\theta, \quad t_2 = t\sin\theta \qquad (0 \leqslant \theta \leqslant 2\pi;\ 0 \leqslant t < \infty), \tag{4.6}$$

and describe one of the rays of our pencil by

$$x_1 = \lambda'_1 t + a_1, \quad x_2 = \lambda'_2 t + a_2, \quad x_3 = \lambda'_3 t + a_3 \qquad (0 \leqslant t < \infty), \tag{4.7}$$

where

$$\begin{aligned} \lambda'_1 &= \lambda^1_1 \cos\theta + \lambda^2_1 \sin\theta, \quad \lambda'_2 = \lambda^1_2 \cos\theta + \lambda^2_2 \sin\theta, \\ \lambda'_3 &= \lambda^1_3 \cos\theta + \lambda^2_3 \sin\theta. \end{aligned} \tag{4.8}$$

Looking at the equations (4.7),(4.8), we see that we have an entire family $(0 \leqslant \theta \leqslant 2\pi)$ of "billiard balls" spreading out within the plane L_2. Assuming the initial point a to be within γ_3, we reflect the billiard balls (4.7) in the facets of γ_3.

Again we let the Euler spline do all reflections for us, and we obtain the parametric equations of the polyhedron

$$\Pi^2 : \begin{aligned} x_1 &= \langle \lambda^1_1 t_1 + \lambda^2_1 t_2 + a_1 \rangle \\ x_2 &= \langle \lambda^1_2 t_1 + \lambda^2_2 t_2 + a_2 \rangle \\ x_3 &= \langle \lambda^1_3 t_1 + \lambda^2_3 t_2 + a_3 \rangle \quad (-\infty < t_1, t_2 < \infty). \end{aligned} \tag{4.9}$$

We call Π^2 *a K-S polyhedron.*

To avoid lower-dimensional situations, we shall usually assume that the coefficients λ^j_i in (4.9) satisfy the conditions

$$\begin{vmatrix} \lambda^1_1 & \lambda^2_1 \\ \lambda^1_2 & \lambda^2_2 \end{vmatrix} \neq 0, \quad \begin{vmatrix} \lambda^1_1 & \lambda^2_1 \\ \lambda^1_3 & \lambda^2_3 \end{vmatrix} \neq 0, \quad \begin{vmatrix} \lambda^1_2 & \lambda^2_2 \\ \lambda^1_3 & \lambda^2_3 \end{vmatrix} \neq 0. \tag{4.10}$$

Equivalently, we can generate Π^2 as follows. Let the plane

$$L_2 : A_1 x_1 + A_2 x_2 + A_3 x_3 = c \qquad (A_1 \neq 0, A_2 \neq 0, A_3 \neq 0) \tag{4.11}$$

intersect γ_3 in a convex polygon bounded by (at most six) sides $s_1, s_2, \ldots, s_r$. We reflect L_2 along each s_i with respect to the facet of γ_3 containing s_i, obtaining r new planes $L^2_1, \ldots, L^2_r$. Each of these r planes intersect γ_3, and are to be reflected just as we reflected L_2. This process of reflection is continued indefinitely. The characteristic structure of a K-S polyhedron is therefore the following: Π^2 *is contained in* γ_3, *and all its edges are in facets of* γ_3, *while adjacent facets of* Π^2 *form equal dihedral angles with the facet of* γ_3 *on which they intersect.*

The simplest example of a Π^2 is obtained if we choose as (4.11) the plane

$$L_2 : x_1 + x_2 + x_3 = 1. \tag{4.12}$$

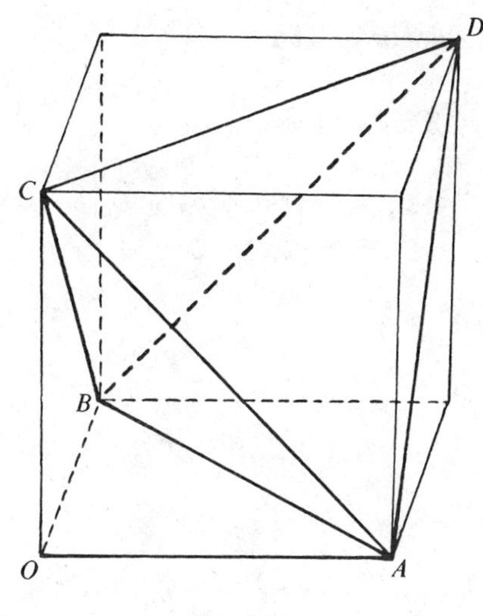

FIG. 15.8

A glance at Figure 15.8 shows that Π^2 is the wonderfully simple *regular tetrahedron* $T = ABCD$. Indeed, notice that T *is closed with respect to reflections in the facets of* γ_3: Any two facets of T, like ABC and ABD, intersect in a diagonal AB of a facet of γ_3 and form equal angles with that facet. I do not know if this remarkable property of T was known before König and Szücs pointed it out on page 87 of [**2**]. The possibility of inscribing a regular tetrahedron in a cube was first noticed by J. Kepler, and we call T *Kepler's tetrahedron*. In our last MTE on b.b. motions we will point out an extremum property of T among all Π^2 satisfying the conditions (4.10).

Observe that we have not used Euler's spline. However, if we represent (4.12) in its parametric form

$$x_1 = t_1,$$
$$x_2 = t_2, \tag{4.13}$$
$$x_3 = 1 - t_1 - t_2,$$

then the equations

$$\Pi^2 = T: \begin{array}{l} x_1 = \langle t_1 \rangle, \\ x_2 = \langle t_2 \rangle, \\ x_3 = \langle 1 - t_1 - t_2 \rangle, \end{array} \tag{4.14}$$

provide a parametric representation of T.

We conclude the present MTE with

5. A Polyhedral Analogue of Theorem 1 (§3).

THEOREM 2. *The* K-S *polyhedron*

$$\Pi^2: \begin{aligned} x_1 &= \langle \lambda_1^1 t_1 + \lambda_1^2 t_2 + a_1 \rangle \\ x_2 &= \langle \lambda_2^1 t_1 + \lambda_2^2 t_2 + a_2 \rangle \\ x_3 &= \langle \lambda_3^1 t_1 + \lambda_3^2 t_2 + a_3 \rangle, \end{aligned} \tag{5.1}$$

such that

$$\textit{the matrix } \|\lambda_i^j\| \textit{ is of rank } 2, \tag{5.2}$$

is closed and finite, if and only if

$$\begin{matrix} \lambda_1^1, & \lambda_2^1, & \lambda_3^1 & \textit{are proportional to three rational numbers and} \\ \lambda_1^2, & \lambda_2^2, & \lambda_3^2 & \textit{are proportional to three rational numbers.} \end{matrix} \tag{5.3}$$

This is an easy consequence of Theorem 1.

(i) *The conditions are sufficient*, for if, after appropriate changes of scale in t_1 and t_2,

$$\lambda_i^j = \frac{1}{q} m_i^j, \tag{5.4}$$

where m_i^j and q are integers, $q > 0$, while all m_i^j and q are relatively prime, then the three functions on the right-hand sides of (5.1) have the period $T = 2q$ in each of the variables t_1 and t_2. It follows that we already obtain the entire Π^2 by restricting t_1 and t_2 to the square

$$0 \leqslant t_1 \leqslant 2q, \quad 0 \leqslant t_2 \leqslant 2q, \tag{5.5}$$

and Π^2 is closed and finite.

(ii) *The conditions are necessary*, for if Π^2 is closed and finite, then the b.b. motion $\Pi^1(t_2)$ obtained from (5.1), *if we keep t_2 fixed*, is certainly closed and finite, and Theorem 1 implies the first condition (5.3). Likewise, keeping t_1 fixed and letting t_2 vary, we obtain the second condition (5.3).

REMARKS. 1. Assuming the stronger conditions (4.10), we can change parameters and replace t_1 and t_2 by

$$t_1' = \lambda_1^1 t_1 + \lambda_1^2 t_2 + a_1, \quad t_2' = \lambda_2^1 t_1 + \lambda_2^2 t_2 + a_2,$$

obtaining

COROLLARY 1. *The* K-S *polyhedron*

$$\Pi^2: \begin{aligned} x_1 &= \langle t_1 \rangle, \\ x_2 &= \langle t_2 \rangle, \\ x_3 &= \langle \alpha t_1 + \beta t_2 + a \rangle \qquad (\alpha \neq 0, \beta \neq 0), \end{aligned} \tag{5.6}$$

is closed and finite if and only if

$$\alpha \text{ and } \beta \text{ are both rational numbers.} \tag{5.7}$$

Notice that the tetrahedron (4.14) is a special case of (5.6).

2. If the condition (5.7) holds, then the plane of every facet of (5.6) is readily seen to have an equation of the form

$$L_2\colon A_1x_1 + A_2x_2 + A_3x_3 = c, \quad \text{all } A_i \text{ are integers}, \quad \text{all } A_i \neq 0. \tag{5.8}$$

Conversely, every initial plane (5.8), which intersects γ_3 strictly, generates by reflections in the facets of γ_3 a closed and finite K-S polyhedron Π^2 of the form (5.6). This is so because we can write (5.8) in the form

$$x_1 = t_1, \quad x_2 = t_2, \quad x_3 = -A_1A_3^{-1}t_1 - A_2A_3^{-1}t_2 + cA_3^{-1}.$$

Problems

1. Following the paper [**2**], our discussion deals mainly with the 3-dimensional case of motions in γ_3. The reader is invited to formulate the corresponding simpler results for b.b. motions in γ_2. Specifically: What are the equations for the polygons of Figure 15.2?

2. Derive the parametric equations for the octagon of Figure 15.3(b), and verify that it is a K-S polygon. *Hint:* See the solution of the problem for the hexagon of Figure 15.3(a) at the beginning of §3.

3. Describe the shape, size and structure of the K-S polyhedron

$$\Pi^2: \begin{aligned} x_1 &= \langle t_1 \rangle, \\ x_2 &= \langle t_2 \rangle, \\ x_3 &= \langle \tfrac{1}{2} - t_1 - t_2 \rangle. \end{aligned} \tag{1}$$

The problem is to show that Π^2 is a closed (self-intersecting) polyhedron having 12 facets, of which 8 are regular triangles and 4 are regular hexagons. For its solution it helps to realize that due to period 2 of each of the functions (1) in each of the two variables t_1 and t_2, it suffices to restrict the variables to the half-closed square

$$S:\quad 0 \leqslant t_1 < 2, \quad 0 \leqslant t_2 < 2.$$

Therefore Π^2 is the image of S into γ_3, by the mapping (1). Since we have to identify in pairs the opposite sides of S, because of the mentioned periodicity, we conclude that Π^2 is topologically a *torus*, or doughnut.

For the reader's enjoyment we now present

A direct construction of Π^2. We do this in a roundabout way by first describing

(i) *Kepler's Stella Octangula.* Our Figure 15.9, in which the 12 edges of the big cube γ_3 are to be disregarded, represents Kepler's beautiful Eight-pointed Star, also called Stella Octangula, which we abbreviate S.O. It is the union of the two Kepler tetrahedra inscribed in γ_3. Another way to obtain the S.O. is as follows.

We dissect γ_3 into 8 small cubes $c_1,\ldots,c_8$, having a common vertex in the center C of γ_3. In each of these cubes c_i we inscribe the Kepler tetrahedron T_i having as one of its vertices the common vertex of c_i and γ_3. The S.O. is the union of these 8 tetrahedra T_i.

(ii) *The construction of* Π^2. Our Figure 15.10 shows the figure obtained by performing on the S.O. of Figure 15.9 the following transformation. We recall that in each cube we can inscribe two Kepler tetrahedra which are symmetric images with respect to the center of the cube. *In each cube* c_i (Figure 15.9) *we replace* T_i *by the second Kepler tetrahedron* T_1' *inscribed in* c_i. This is all, because we claim that

$$\Pi^2 = \bigcup_{i=1}^{8} T_i'. \tag{2}$$

Proof of (2). In the equations (1) we restrict the variables t_1, t_2, to the triangle

$$\Delta:\quad 0 \leqslant t_1, \quad 0 \leqslant t_2, \quad t_1 + t_2 \leqslant \tfrac{1}{2} \tag{3}$$

and wish to find its image by (1) into γ_3. From properties of the Euler spline we see that in Δ we have $x_1 = t_1, x_2 = t_2, x_3 = \frac{1}{2} - t_1 - t_2$, and therefore $x_1 + x_2 + x_3 = \frac{1}{2}$. This shows that Δ is mapped into the plane

$$L_2: x_1 + x_2 + x_3 = \tfrac{1}{2} \tag{4}$$

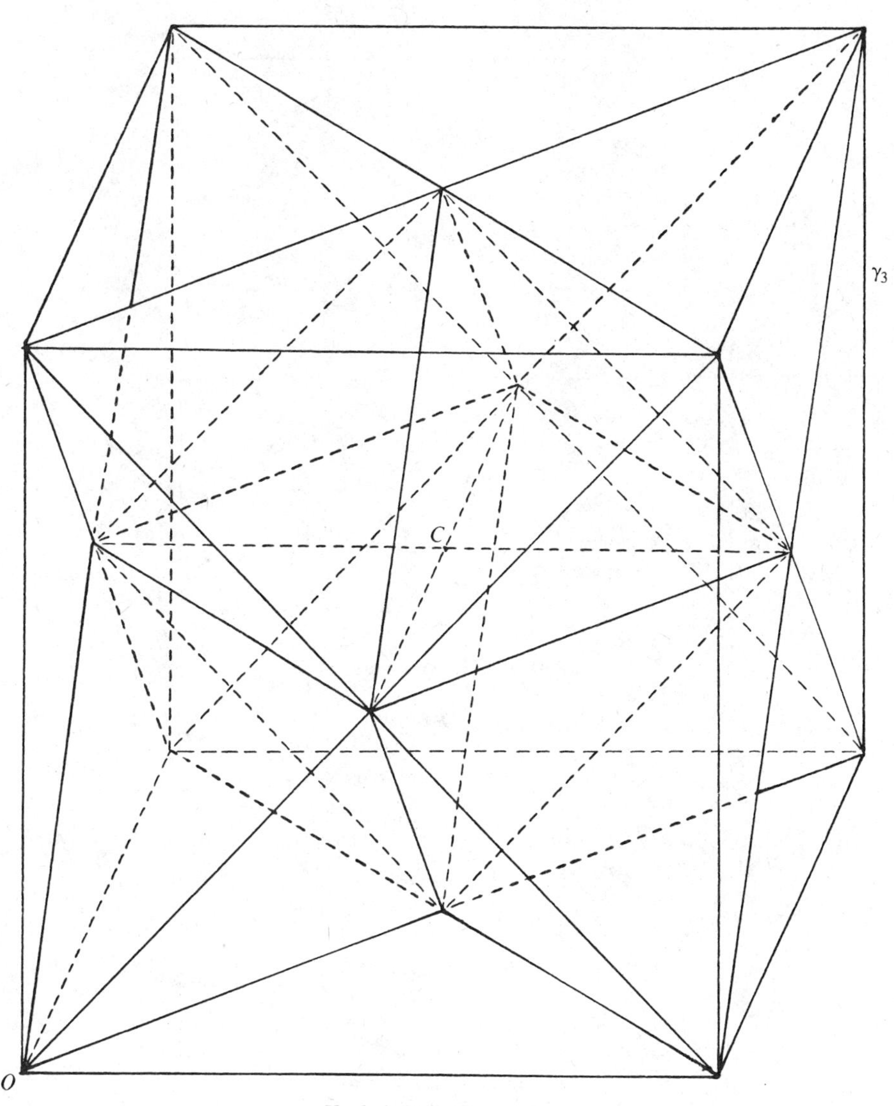

Kepler's Stella Octangula.

FIG. 15.9

and that one facet of Π^2 is

$$\tau_1 = \text{triangle } abc = L_2 \cap \gamma_3, \tag{5}$$

marked in Figure 15.10. We conclude: *The polyhedron Π^2 arises by reflections of the plane L_2 in the facets of γ_3*. This conclusion allows us to establish (2).

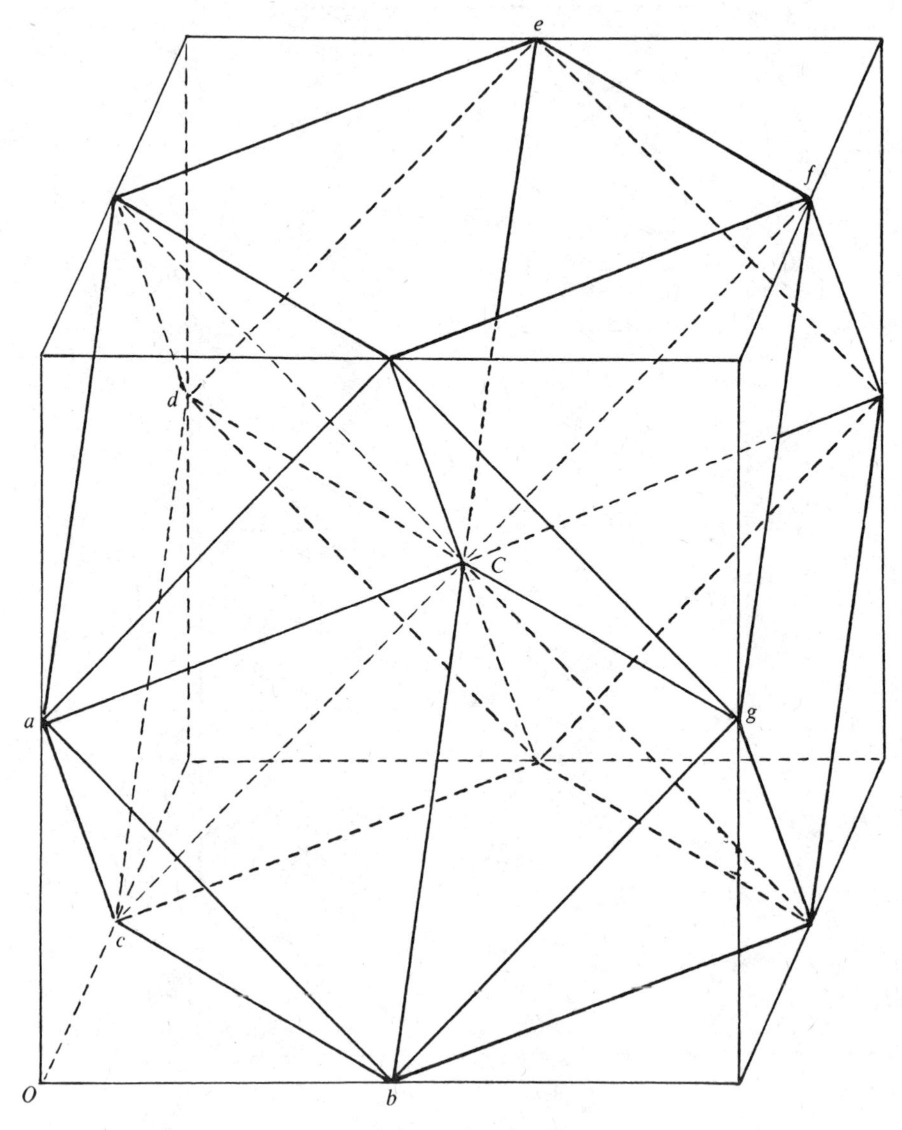

The K-S polyhedron Π^2.

FIG. 15.10

To do this, we observe first that in Figure 15.10 the 6 points b, c, d, e, f, g, are in a plane π_1 through C, and that they are the vertices of a regular hexagon H_1, with center at C. Because the triangle bcC is the reflection of abc along their common side bc, it follows that

$$H_1 = \pi_1 \cap \gamma_3 \tag{6}$$

is the complete reflection of L_2, within γ_3, along the segment bc.

Observe now the following: Consider the center C as the *the vertex* of the 8 tetrahedra T_i'. They have a total of $3\times 8 = 24$ *lateral facets*. These $24 = 4\times 6$ lateral facets give rise to 4 regular hexagons H_1, H_2, H_3, H_4, among which the hexagon (6) is the first. If τ_i denotes the base of T_i', then clearly

$$\bigcup_{i=1}^{8} T_i' = \left(\bigcup_{i=1}^{8} \tau_i\right) \cup \left(\bigcup_{i=1}^{4} H_j\right). \tag{7}$$

The system of 12 planes on the right-hand side is closed with respect to reflections in the facets of γ_3; since L_2 is one of the planes, by (4) and (5), we conclude that Π^2 is identical with the right-hand side of (7), and (2) is thereby established.

The reader is urged to make a model of Π^2. For this he needs 8 tetrahedra T_i' (made of good paper) which are joined together along the 12 segments joining C to the midpoints of the edges of γ_3.

The polyhedron Π^2 of Figure 15.10 was introduced in [**2**, p. 88]. It is a 3-dimensional analogue of the square of Figure 15.2(c).

4. Show that the K-S polyhedron

$$\Pi^2(a): \begin{aligned} x_1 &= \langle t_1 \rangle \\ x_2 &= \langle t_2 \rangle \\ x_3 &= \langle a - t_1 - t_2 \rangle \qquad (a \text{ fixed}, 0 < a < 1), \end{aligned} \tag{8}$$

has a surface area $A(a)$ which does not depend on a and is equal to

$$A(a) = 4\sqrt{3}\,. \tag{9}$$

Solution. Divide the square

$$S:\quad 0 \leqslant t_1 \leqslant 2, \quad 0 \leqslant t_2 \leqslant 2$$

into the 12 polygons $p_1, p_2, \ldots, p_{12}$, (8 triangles and 4 hexagons) in which none of the functions

$$t_1, t_2, a - t_1 - t_2$$

assume integral values. The polygon p_j ($j=1,\ldots,12$) is mapped by (8) into a facet of $\Pi^2(a)$ whose area is given by

$$A_j = \iint_{p_j} \sqrt{EG - F^2}\, dt_1\, dt_2, \tag{10}$$

where $ds^2 = E(dt_1)^2 + 2F\,dt_1\,dt_2 + G(dt_2)^2$ is the square of the line element of $\Pi^2(a)$ in p_j (see [5, p. 80]). From (8) we find in each of the p_j that $dx_1 = \pm dt_1$, $dx_2 = \pm dt_2$, $dx_3 = \pm(dt_1 + dt_2)$, and therefore

$$ds^2 = (dx_1)^2 + (dx_2)^2 + (dx_3)^2 = 2(dt_1)^2 + 2\,dt_1\,dt_2 + 2(dt_2)^2.$$

This shows that $E=2$, $F=1$, $G=2$, and therefore by (10) $A_j = \sqrt{3}$ Area p_j. Now

$$S(a) = \sum_{j=1}^{12} A_j = \sqrt{3} \sum \text{Area}\, p_j = \sqrt{3}\ \text{Area}\ S = 4\sqrt{3}$$

and (9) is established.

If $a \to 0$, or $a \to 1$, then $\Pi^2(a)$ collapses and reduces to the Kepler tetrahedron T, defined by (4.13) of §4, *covered twice*, and the surface area of T is $2\sqrt{3}$.

References

[1] G. H. Hardy and E. M. Wright, *An Introduction to the Theory of Numbers*, 3rd ed., Oxford, 1954.

[2] D. König and A. Szücs, *Mouvement d'un point abandonné à l'intérieur d'un cube*, Rend. Cir. Mat. Palermo (2), 36 (1913) 79–90.

[3] F. Lettenmeyer, *Neuer Beweis des allgemeinen Kroneckerschen Approximationssatzes*, Proc. London Math. Soc. (2), 21 (1922) 306–314.

[4] I. J. Schoenberg, *Extremum problems for the motions of a billiard ball* II. *The L_∞ norm*, Indag. Math., 38 (1976) 263–279.

[5] J. J. Stoker, *Differential Geometry*, Wiley-Interscience, New York, 1969.

CHAPTER **16**

ON THE MOTIONS OF A BILLIARD BALL II: THE THEOREMS OF KRONECKER

1. Introduction. The four chapters, 15, 16, 17, and 18, on billiard ball motions are related as follows. In Chapter 15 we introduced the König-Szücs polygons and polyhedra. These are very special polygons and polyhedra having remarkable properties. Their global arithmetical behavior is presented in Chapter 17 and is based on the work of L. Kronecker (1823–1891), belonging to a field known as Diophantine Approximations. In the present chapter we develop some of Kronecker's results following the ideas of F. Lettenmeyer from his paper [**2**]. These results are here presented without references to the other chapters, and may therefore be read independently. For references to the closely related ideas of Hermann Weyl (1885–1955) on Uniform Distribution mod 1, we refer to the books of Hardy–Wright [**1**, §23.10] and I. Niven [**3**, Chapter 6, §§3 and 4].

Also Chapter 18 may be read independently, being devoted to a geometric extremum property of a particular K-S polyhedron known as Kepler's Stella Octangula.

2. The One-dimensional Kronecker Theorem. We discuss this simplest of Kronecker's theorems for two reasons: 1. It is likely the most important among his results. 2. It already exhibits Lettenmeyer's idea also used in proving the higher-dimensional theorems.

THEOREM 1 (Kronecker). *If θ is irrational, then the set*

$$\sum = \{n\theta + m;\ n \geqslant 1 \text{ and } m \text{ are integers}\} \tag{2.1}$$

is dense on the real axis.

By "dense on $\mathbb{R}$" we mean that every neighborhood of any point of $\mathbb{R}$ contains a point of Σ.

Proof: We say that $a \equiv b \pmod 1$, provided that $a-b$ is an integral multiple of 1, hence an integer. Writing

$$A_n = n\theta, \qquad (n=1,2,\ldots), \tag{2.2}$$

we denote by $A'_n = n\theta + m$ an unspecified number such that $A'_n \equiv A_n$ (mod 1). Clearly Σ is the set of all the A'_n.

Notice that all points of Σ are distinct, in particular

$$A'_n \neq A'_m \quad \text{if} \quad n \neq m; \tag{2.3}$$

because of the irrationality of θ.

The set Σ has the following *additive property*:

$$A'_r + A'_s = A'_{r+s} \quad \textit{and} \quad A'_r - A'_s = A'_{r-s} \quad \textit{if} \quad r > s, \tag{2.4}$$

because if $A'_r = r\theta + r_1$, $A'_s = s\theta + s_1$, then $A'_r + A'_s = (r+s)\theta + r_1 + s_1 = A'_{r+s}$, and $A'_r - A'_s = (r-s)\theta + r_1 - s_1 = A'_{r-s}$.

As usual, we denote by $[x]$ the largest integer not exceeding x, and then

$$(x) = x - [x] \tag{2.5}$$

is the fractional part of x. Let

$$\Pi = \{(n\theta);\, n = 1,2,\ldots\} \tag{2.6}$$

denote the set of fractional parts of the numbers (2.2). Clearly

$$\Pi = \Sigma \cap [0,1). \tag{2.7}$$

All elements of Π are distinct because of the irrationality of θ and the set Π is therefore an infinite set. Since Π is also a bounded set, the Bolzano-Weierstrass theorem implies that Π has a limit point l. Given $\delta > 0$, we can therefore find in the interval $(l - \delta/2, l + \delta/2)$ a pair of points A'_r, A'_s such that

$$r > s. \tag{2.8}$$

We now form the equidistant sequence of points

$$k(A'_r - A'_s) \qquad (k = 1,2,\ldots). \tag{2.9}$$

By the additive property (2.4) and our assumption (2.8), the point (2.9) belongs to Σ, in fact it is $= A'_{kr-ks}$. Since $|A'_r - A'_s| < \delta$, we conclude that the points (2.9) are within a distance $< \delta$ from every point of $[0, \infty)$, or of $(-\infty, 0]$, depending on the sign of the step $A'_r - A'_s$. Since translations by integers of the entire sequence (2.9) will keep it within Σ, and since δ is arbitrarily small, the theorem is established.

From Theorem 1 and (2.7) we conclude that

$$\textit{the set } \Pi\textit{, defined by } (2.6)\textit{, is dense in the interval } [0,1]. \tag{2.10}$$

3. The First Theorem of Kronecker in $\mathbb{R}^3$. Theorem 1 of §2 generalizes to higher dimensions. We should first consider a generalization to $\mathbb{R}^2$, but find it convenient to pass to $\mathbb{R}^3$. We need a few definitions.

1. Let $A=(a_1, a_2, a_3)$ and $B=(b_1, b_2, b_3)$ be two points of $\mathbb{R}^3$. *We write*

$$A \equiv B \pmod 1 \tag{3.1}$$

meaning thereby that the 3 *differences* $a_i - b_i$ *are integers.*

2. That the number θ, of §2, is irrational may be expressed as follows: A relation $C_1\theta + C_2 = 0$ with integer C_1 and C_2, implies that $C_1 = 0$ and $C_2 = 0$. This generalizes nicely as follows:

We say that the reals $a_1, a_2, \ldots, a_k$ *are linearly independent* (abbreviated to lin. ind.) *provided that a linear equation* $C_1a_1 + C_2a_2 + \cdots + C_ka_k = 0$ *with integer coefficients* C_i, *implies that all* C_i *are* 0. (3.2)

An example: If $k=3$ and $a_1 = \log 2$, $a_2 = \log 3$, $a_3 = \log 5$, then (3.2) becomes $C_1\log 2 + C_2 \log 3 + C_3 \log 5 = 0$, and this may be written as $2^{C_1}3^{C_2}5^{C_3} = 1$. However, the fundamental theorem of arithmetic implies that $C_1 = C_2 = C_3 = 0$.

Let a_1, a_2, a_3 be three reals such that

$$a_1, a_2, a_3, 1 \textit{ are linearly independent.} \tag{3.3}$$

We are concerned with the sequence of equidistant points

$$A_n = (na_1, na_2, na_3) \qquad (n=1,2,\ldots). \tag{3.4}$$

However, we consider also all points of $\mathbb{R}^3$ which are congruent to these points mod 1, according to the definition of (3.1). In particular, we denote by A'_n an *unspecified* point such that

$$A'_n \equiv A_n \pmod 1, \tag{3.5}$$

hence such that $A'_n = (na_i + n_i)$ with integer n_i. Let $\{A'_n\}$ denote the set of A'_n satisfying (3.5), where it is good to visualize $\{A'_n\}$ as the 3-dimensional lattice of points of which A_n is one of its points.

Finally, *let*

$$\Sigma = \bigcup_{n=1}^{\infty} \{A'_n\} \tag{3.6}$$

denote the set of points of $\mathbb{R}^3$ *which are congruent* mod 1 *to a point of the sequence* (3.4).

Notice that *all points of* Σ *are distinct, in particular*

$$A'_n \neq A'_m \quad \text{if} \quad n \neq m. \tag{3.7}$$

Indeed, the equations $na_i + n_i = ma_i + m_i$ $(i=1,2,3)$ and $n \neq m$, would imply that all a_i are rational, contradicting (3.3).

We can now state Kronecker's first theorem in $\mathbb{R}^3$ as

THEOREM 2 (Kronecker). *The set* Σ *defined by* (3.6), (3.5), *and* (3.4) *is dense in the entire space* $\mathbb{R}^3$.

Proof: Using the fractional part function (2.5), we consider also the sequence of points

$$P_n = ((na_1), (na_2), (na_3)) \qquad (n=1,2,\ldots). \tag{3.8}$$

In analogy with the 1-dimensional case, we could call P_n the least positive residue mod 1 of the point (3.4). As in (2.6), let

$$\Pi = \{P_n; n=1,2,\ldots\}. \tag{3.9}$$

The analogue of (2.7) is the set relation

$$\Pi = \Sigma \cap \gamma_3, \tag{3.10}$$

where

$$\gamma_3 = \{0 \leq x_i < 1; i=1,2,3\} \tag{3.11}$$

is the unit cube.

An equivalent formulation of Theorem 2 is

THEOREM 3. *The set* Π, *defined by* (3.9) *and* (3.8), *is dense in the unit cube* γ_3.

That Theorem 2 implies Theorem 3 should be clear from (3.10). For the converse: If we translate γ_3 by the vector (n_1, n_2, n_3), then γ_3 becomes the cube

$$C(n_1, n_2, n_3) = \{(x_1, x_2, x_3); n_i \leq x_i < n_i + 1, i=1,2,3\}, \tag{3.12}$$

and Π is translated into $\Sigma \cap C(n_1, n_2, n_3)$. Hence Σ is dense in every cube

(3.12), and therefore also in $\mathbb{R}^3$. Even though Theorems 2 and 3 are equivalent, Theorem 2 is a little easier to establish.

Identifying the point A'_n with the vector $\overrightarrow{OA'_n}$, the set (3.6) enjoys the following *additive property*:

$$A'_r + A'_s = A'_{r+s}, \quad \textit{and also} \quad A'_r - A'_s = A'_{r-s} \quad \textit{if} \quad r > s. \quad (3.13)$$

Indeed, if $A'_r = (ra_i + r_i)$, $A'_s = (sa_i + s_i)$, then

$$A'_r + A'_s = ((r+s)a_i + r_i + s_i) = A'_{r+s}$$

and

$$A'_r - A'_s = ((r-s)a_i + r_i - s_i) = A'_{r-s}.$$

The set Π is an infinite bounded set, and again by the Bolzano-Weierstrass theorem it has a limit point l in the closure of γ_3. Given $\delta > 0$ we can therefore find points P_r and P_s such that

$$|P_r - l| < \delta, \quad |P_s - l| < \delta, \qquad r > s \quad (3.14)$$

where

r and s are independently arbitrary large with $r > s$. (3.15)

We now form the vector $\overrightarrow{P_sP_r} = P_r - P_s$ and translate it so as to make it start at the point $P_1 = (a_1, a_2, a_3)$, where *without loss of generality we may assume that**

$$0 < a_1 < 1, \quad 0 < a_2 < 1, \quad 0 < a_3 < 1. \quad (3.16)$$

We may even impose on δ the additional condition that

$$\delta < \tfrac{1}{2}\min(a_1, a_2, a_3, 1 - a_1, 1 - a_2, 1 - a_3). \quad (3.17)$$

Applying the additive property (3.13) we conclude that

$$P_1 + \overrightarrow{P_sP_r} = P_1 + P_r - P_s = P_1 + A'_{r-s} = A'_{r-s+1}. \quad (3.18)$$

We already know from our definitions that we have the *congruence*

$$A'_{r-s+1} \equiv P_{r-s+1} \pmod 1. \quad (3.19)$$

---

*See the Remark at the end of this section.

Let us show that we actually have the *equality*

$$A'_{r-s+1} = P_{r-s+1}. \tag{3.20}$$

From (3.14) we obtain that

$$|\overrightarrow{P_sP_r}| < 2\delta. \tag{3.21}$$

Writing $A'_{r-s+1} = (x_1, x_2, x_3)$, then from (3.21) and (3.17) we find

$$0 < a_i - 2\delta < x_i < a_i + 2\delta < 1 \qquad (i=1,2,3)$$

and therefore $0 < x_i < 1$ $(i=1,2,3)$. This forces the congruence (3.19) to become the equality (3.20). Our final result is that

$$P_1 + \overrightarrow{P_sP_r} = P_{r-s+1}, \tag{3.22}$$

and shows that

$$|P_{r-s+1} - P_1| < 2\delta. \tag{3.23}$$

By (3.14) and (3.15) *the subscript* $r-s+1$ *of* P_{r-s+1} *may assume arbitrarily large values*. It follows that within the sphere

$$S_\delta : |x - P_1| < 2\delta \tag{3.24}$$

the set Π has infinitely many points. Let this set be

$$\sigma = \Pi \cap S_\delta = \bigcup_{r \in N_\delta} \{P_r\} \text{ where } N_\delta \text{ is an infinite set.} \tag{3.25}$$

Two cases are a priori possible.

Case 1. The point-set σ is contained in a plane π passing through P_1.

Case 2. The set σ is not contained in a plane through P_1.

We claim that

$$\textit{Case } 1 \textit{ cannot occur.} \tag{3.26}$$

Proof: We assume that Case 1 occurs and wish to get a contradiction. Let r_1, r_2, r_3 be increasing elements of the set N_δ, $1 < r_1 < r_2 < r_3$, so that $P_1, P_{r_1}, P_{r_2}, P_{r_3}$, are four points of the plane π. This implies that their coordinates satisfy the equation

$$\begin{vmatrix} a_1 & a_2 & a_3 & 1 \\ (r_1a_1) & (r_1a_2) & (r_1a_3) & 1 \\ (r_2a_1) & (r_2a_2) & (r_2a_3) & 1 \\ (r_3a_1) & (r_3a_2) & (r_3a_3) & 1 \end{vmatrix} = 0.$$

This we rewrite as

$$\begin{vmatrix} a_1 & a_2 & a_3 & 1 \\ r_1a_1-[r_1a_1] & r_1a_2-[r_1a_2] & r_1a_3-[r_1a_3] & 1 \\ r_2a_1-[r_2a_1] & r_2a_2-[r_2a_2] & r_2a_3-[r_2a_3] & 1 \\ r_3a_1-[r_3a_1] & r_3a_2-[r_3a_2] & r_3a_3-[r_3a_3] & 1 \end{vmatrix} = 0.$$

Multiplying the first row successively by r_1, r_2, r_3 and subtracting from the second, third, and fourth row, respectively, we obtain the equation

$$\begin{vmatrix} a_1 & a_2 & a_3 & 1 \\ [r_1a_1] & [r_1a_2] & [r_1a_3] & r_1-1 \\ [r_2a_1] & [r_2a_2] & [r_2a_3] & r_2-1 \\ [r_3a_1] & [r_3a_2] & [r_3a_3] & r_3-1 \end{vmatrix} = 0. \tag{3.27}$$

Observe that the elements in the last three rows are integers. Expanding the determinant by the elements of the first row, we find that (3.27) represents a linear equation of the form $C_1a_1 + C_2a_2 + C_3a_3 + C_4 = 0$, with integer coefficients. By our assumption (3.3), these coefficients must vanish, and in particular $C_1 = 0$. From (3.27) we therefore obtain the equation

$$\begin{vmatrix} [r_1a_2] & [r_1a_3] & r_1-1 \\ [r_2a_2] & [r_2a_3] & r_2-1 \\ [r_3a_2] & [r_3a_3] & r_3-1 \end{vmatrix} = 0 \qquad (1<r_1<r_2<r_3).$$

On dividing the last row by $r_3 - 1$ we obtain

$$\begin{vmatrix} [r_1a_2] & [r_1a_3] & r_1-1 \\ [r_2a_2] & [r_2a_3] & r_2-1 \\ \dfrac{[r_3a_2]}{r_3-1} & \dfrac{[r_3a_3]}{r_3-1} & 1 \end{vmatrix} = 0.$$

Letting here $r_3 \to \infty$ we obtain in the limit the equation

$$\begin{vmatrix} [r_1 a_2] & [r_1 a_3] & r_1 - 1 \\ [r_2 a_2] & [r_2 a_3] & r_2 - 1 \\ a_2 & a_3 & 1 \end{vmatrix} = 0.$$

This is a linear equation $C_2 a_2 + C_3 a_3 + C_4 = 0$, and again (3.3) implies that $C_2 = 0$, hence that

$$\begin{vmatrix} [r_1 a_3] & r_1 - 1 \\ [r_2 a_3] & r_2 - 1 \end{vmatrix} = 0.$$

This shows that

$$\begin{vmatrix} [r_1 a_3] & r_1 - 1 \\ \dfrac{[r_2 a_3]}{r_2 - 1} & 1 \end{vmatrix} = 0,$$

and letting $r_2 \to \infty$ we obtain that

$$\begin{vmatrix} [r_1 a_3] & r_1 - 1 \\ a_3 & 1 \end{vmatrix} = 0.$$

However, this equation shows that a_3 *is a rotational number* in contradiction to our assumption (3.3).

We have just established (3.26) *and therefore Case* 2 *is the only possibility.*

Accordingly, we can find within the sphere (3.24) three points $P_{r_1}, P_{r_2}, P_{r_3}$ $(1 < r_1 < r_2 < r_3)$, such that the points

$$P_1, P_{r_1}, P_{r_2}, P_{r_3} \textit{ are the vertices of a nondegenerate tetrahedron}. \quad (3.28)$$

A proof of Theorem 2 is now obtained as follows:

1. Because of (3.28) we can take the vectors $\overrightarrow{P_1 P_{r_1}}, \overrightarrow{P_1 P_{r_2}}, \overrightarrow{P_1 P_{r_3}}$ as the unit vectors of an oblique coordinate system with origin at P_1, and consider the oblique lattice of points

$$P_1 + n_1 \overrightarrow{P_1 P_{r_1}} + n_2 \overrightarrow{P_1 P_{r_2}} + n_3 \overrightarrow{P_1 P_{r_3}}, \qquad \text{all } n_i \text{ integers} \geqslant 0. \quad (3.29)$$

By repeated application of the property (3.13) we conclude that all the points (3.29) belong to the set Σ.

2. All the edges of the tetrahedron (3.28) are $< 2\delta$, and this implies that the points (3.29) are within a distance 4δ from every point of the positive octant of our oblique coordinate system.

3. The set (3.29) may be translated by an arbitrary vector of integer components, thereby assuming positions which are all within Σ. This shows that every point of $\mathbb{R}^3$ is within a distance 4δ from some point of Σ.

4. Finally, δ was arbitrarily small, and therefore Σ is dense in $\mathbb{R}^3$, proving Theorem 2.

Our Theorems 2 and 3 generalize immediately to

THEOREM 4. *We assume that*

$$a_1, a_2, a_3, 1 \text{ are linearly independent}, \tag{3.30}$$

and consider the sequence of points

$$A_n = (na_1 + c_1, na_2 + c_2, na_3 + c_3) \qquad (n = 1, 2, \ldots), \tag{3.31}$$

where the c_i are reals. Then the set Σ_c of points of R^3 which are congruent (mod 1) *to points of the sequence* (3.31), *are dense in $\mathbb{R}^3$.*

Proof: It suffices to observe that the set Σ_c is obtained from the old set (3.6) by a translation by the vector (c_1, c_2, c_3), as indicated by

$$\sum_c = \sum + (c_1, c_2, c_3). \tag{3.32}$$

Clearly Σ_c is dense in $\mathbb{R}^3$.

We may also state

THEOREM 5. *We again assume that* (3.30) *holds and consider the least positive residues*

$$P_n = ((na_1 + c_1), (na_2 + c_2), (na_3 + c_3)) \qquad (n = 1, 2, \ldots) \tag{3.33}$$

of the points (3.31), *then the set*

$$\Pi_c = \{P_n; n = 1, 2, \ldots\} \tag{3.34}$$

is dense in the unit cube γ_3.

Proof: It suffices to observe that

$$\Pi_c = \sum_c \cap \gamma_3. \tag{3.35}$$

REMARK. Let us justify the additional restrictions (3.16) on the numbers a_1, a_2, a_3, of Theorem 2. We already know that all a_i are irrational. *If we* *replace the a_i by their least positive residues (a_i), then the sets Σ and Π remain thereby unchanged*, while the (a_i) clearly satisfy (3.16).

4. The Second Theorem of Kronecker in $\mathbb{R}^3$. As in §3 we consider the sequence of points

$$A_n = (na_1, na_2, na_3) \qquad (n=1,2,\dots), \tag{4.1}$$

and define the set Σ as before: We denote by A'_n an unspecified point $\equiv A_n$ (mod 1). They form a 3-dimensional lattice $\{A'_n\}$, and

$$\Sigma = \bigcup_1^\infty \{A'_n\}. \tag{4.2}$$

In contradistinction to §3 we drop the assumption (3.3), and assume instead that there is a linear equation

$$B_1a_1 + B_2a_2 + B_3a_3 + B_4 = 0 \textit{ with integer coefficients}, \tag{4.3}$$

such that

$$\sum_1^3 B_i^2 > 0 \qquad (B_1, B_2, B_3, B_4) = 1. \tag{4.4}$$

We also assume that

> *there is no second equation of the form* (4.3) *which is linearly independent from* (4.3). (4.5)

As a third assumption we assume that

$$a_1, a_2, a_3 \textit{ are three irrational numbers}. \tag{4.6}$$

This is not an essential assumption. However, if a_1 is rational, say, then the equation (4.3) must reduce to the form

$$B_1a_1 + B_4 = 0 \quad \text{with} \quad B_1 \neq 0 \quad \text{and} \quad (B_1, B_4) = 1. \tag{4.7}$$

By (4.5) it follows that the numbers $a_2, a_3, 1$, are linearly independent, because a relation $C_2a_2 + C_3a_3 + C_4 = 0$ would be linearly independent of (4.7). This would reduce our problem to a 2-dimensional one. To avoid this reduction we assume (4.6) to hold. Moreover, in view of the remark at the end of §3, we may also assume, without loss of generality, that

$$0 < a_1 < 1, \quad 0 < a_2 < 1, \quad 0 < a_3 < 1. \tag{4.8}$$

Again, as in (3.7), all points of Σ are distinct, in particular

$$A'_n \neq A'_m \quad \text{if} \quad n \neq m, \tag{4.9}$$

for $A'_n = A'_m$ would imply that all a_i are rational in contradiction to our assumption (4.6).

What can we now say about the location of the set Σ?

This question will be answered in Theorem 4 below. To prepare its statement we observe the following. If

$$A'_n = (x_1, x_2, x_3), \quad \text{where} \quad x_i = na_i + n_i, \tag{4.10}$$

then

$$\sum_1^3 B_i x_i = \sum_1^3 B_i(na_i + n_i) = n\sum_1^3 B_i a_i + \sum_1^3 B_i n_i.$$

From (4.3) we have $\Sigma_1^3 B_i a_i = -B_4$ and so

$$\sum_1^3 B_i x_i = \sum_1^3 B_i n_i - nB_4. \tag{4.11}$$

Since the right-hand side is an integer N, say, we have established

LEMMA 1. *The set* Σ *is contained in the sequence of parallel and equidistant planes*

$$\pi(N): B_1 x_1 + B_2 x_2 + B_3 x_3 = N, \quad \textit{where} \quad N \in \mathbb{Z}. \tag{4.12}$$

We call these the *critical planes* of our problem.

We may now state

THEOREM 6 (Kronecker). *The set* Σ *is dense in the union*

$$\bigcup_{N=-\infty}^{\infty} \pi(N) \tag{4.13}$$

of the critical planes.

This will be established in §6.

5. A Few Lemmas.

LEMMA 2. *Each of the critical planes* (4.12) *contains some point* A'_n *of* Σ.

Proof: We use the equation (4.11), which was implied by (4.3), and wish to determine the integers n and n_i, such that

$$\sum_1^3 B_i n_i - nB_4 = N \qquad (n \geqslant 1),$$

where N is preassigned. This is always possible because of $(B_1, B_2, B_3, B_4) = 1$, and our lemma is established.

LEMMA 3. *If the set* Σ *is dense in the particular critical plane*

$$\pi(\tilde{N}): \sum_1^3 B_i x_i = \tilde{N}, \tag{5.1}$$

then Σ *is dense in every critical plane* $\pi(N)$.

Proof: Let us show that Σ is dense in $\pi(N)$. By Lemma 2 let

$$A'_n \in \pi(N), \quad A'_m \in \pi(\tilde{N}). \tag{5.2}$$

By assumption $\pi(\tilde{N})$ contains points

$$A'_r \in \pi(\tilde{N}) \tag{5.3}$$

such that their set

$$\alpha = \{A'_r\} \quad \textit{is dense in} \quad \pi(\tilde{N}). \tag{5.4}$$

Evidently, the set α contains points A'_r with arbitrarily large subscripts r, for the set of points congruent (mod 1) to a finite number of points $A_1, A_2, \ldots, A_k$ could not possibly satisfy the assumption.

If we translate $\pi(\tilde{N})$ by the vector $\overrightarrow{A'_m A'_n}$, then

1. The plane $\pi(\tilde{N})$ assumes the new position $\pi(N)$, because of (5.2).

2. The new positions of the points of α are clearly dense in $\pi(N)$, by (5.4), while their new positions are

$$A'_r + \overrightarrow{A'_m A'_n} = A'_r + A'_n - A'_m = A'_{r+n-m}$$

and these belong to Σ if $r > m - n$. The new set

$$\alpha + \overrightarrow{A'_m A'_n}$$

is therefore dense in $\pi(N)$.

6. The Crucial Lemma 4. In view of Lemma 3 our Theorem 4 is proved as soon as we establish

LEMMA 4. *The set* $\Sigma = \cup\{A'_n\}$ *is dense in a certain critical plane*

$$\pi(N): \sum_1^3 B_i x_i = N. \tag{6.1}$$

Proof: As in §3 we consider the sequence of points

$$P_n = ((na_1), (na_2), (na_3)) \qquad (n=1,2,\ldots), \tag{6.2}$$

and their set

$$\Pi = \Sigma \cap \gamma_3. \tag{6.3}$$

The point $P_1 = (a_1, a_2, a_3)$ is, by Lemma 1, in some critical plane, and let this plane be $\pi(N)$, defined by (6.1). As in §3, the bounded infinite set Π has a limit point l, $l \in \bar{\gamma}_3$. A sphere $|x-l|<\delta$ contains points P_r and P_s, where r and s are arbitrarily large, $r>s$. We now shift the vector $\overrightarrow{P_sP_r}$ to make it start at the point P_1, obtaining the point

$$P_1 + \overrightarrow{P_sP_r} = A'_{r-s+1}.$$

Because of $|P_r - P_s| < 2\delta$, we conclude that the sphere

$$S_\delta : |x - P_1| < 2\delta$$

contains infinitely many points A'_r of Σ, if $r \in N_\delta$, where N_δ is an infinite set of integers. We now restrict δ further by the two requirements

1. *Let* δ *satisfy*

$$2\delta < \min(a_1, a_2, a_3, 1-a_1, 1-a_2, 1-a_3).$$

As in (3.20), this implies that $A'_r = P_r$ if $r \in N_\delta$.

2. *That* δ *is less than the distance from* P_1 *to the nearest critical plane different from* $\pi(N)$.

It follows that

$$P_r \in \pi(N) \quad \text{if} \quad r \in N_\delta. \tag{6.4}$$

A proof of Lemma 4 uses arguments already used in §3. Concerning the set

$$\sigma = \{P_1\} \cup \{P_r;\, r \in N_\delta\} \tag{6.5}$$

there are two possibilities:

Case 1. The set σ is contained in a straight line λ.

Case 2. The set σ is not contained in a straight line.

We claim that

$$\textit{Case } 1 \textit{ cannot occur.} \tag{6.6}$$

Proof: We assume that the set σ is on a line $\lambda \subset \pi(N)$, and wish to get a contradiction. It follows that if

$$1 < r_1 < r_2, \quad r_1 \in N_8, \quad r_2 \in N_8, \tag{6.7}$$

then

$$\textit{the three points } P_1, P_{r_1}, P_{r_2} \textit{ are collinear.} \tag{6.8}$$

Let us assume that in our basic equation (4.3) we have

$$B_1 \neq 0. \tag{6.9}$$

We project orthogonally the points (6.8) onto the plane $x_1 = 0$ to obtain the points having coordinates

$$(a_2, a_3), \quad ((r_1 a_2), (r_1 a_3)), \quad ((r_2 a_2), (r_2 a_3)).$$

These points being collinear by (6.8), we obtain the equation

$$\begin{vmatrix} a_2 & a_3 & 1 \\ (r_1 a_2) & (r_1 a_3) & 1 \\ (r_2 a_2) & (r_2 a_3) & 1 \end{vmatrix} = 0, \tag{6.10}$$

or

$$\begin{vmatrix} a_2 & a_3 & 1 \\ r_1 a_2 - [r_1 a_2] & r_1 a_3 - [r_1 a_3] & 1 \\ r_2 a_2 - [r_2 a_2] & r_2 a_3 - [r_2 a_3] & 1 \end{vmatrix} = 0.$$

Multiplying the first row successively by r_1 and r_2, and subtracting it from the second and third row, respectively, we obtain

$$\begin{vmatrix} a_2 & a_3 & 1 \\ [r_1 a_2] & [r_1 a_3] & r_1 - 1 \\ [r_2 a_2] & [r_2 a_3] & r_2 - 1 \end{vmatrix} = 0.$$

This is an equation $C_2a_2 + C_3a_3 + C_4 = 0$ with integer coefficients. From (6.9) we see that this would contradict our assumption (4.5), unless $C_2 = C_3 = C_4 = 0$. From $C_2 = 0$ we obtain that

$$\begin{vmatrix} [r_1a_3] & r_1 - 1 \\ [r_2a_3] & r_2 - 1 \end{vmatrix} = 0.$$

Dividing the second row by r_2 and letting $r_2 \to \infty$ (within the infinite set N_8!) we obtain that

$$\begin{vmatrix} [r_1a_3] & r_1 - 1 \\ a_3 & 1 \end{vmatrix} = 0.$$

This, however, is the desired contradiction, because it shows that a_3 is a rational number. This establishes (6.6).

We can now conclude a proof of Lemma 4 by showing that Σ *is dense in the plane* (6.1). Being in Case 2, we can find in the set σ, of (6.4), three points such that

$$P_1, P_{r_1}, P_{r_2} \textit{ are the vertices of a nondegenerate triangle.} \tag{6.11}$$

We take $\overrightarrow{P_1P_{r_1}}$ and $\overrightarrow{P_1P_{r_2}}$ as the unit vectors of oblique coordinates with origin P_1, in the plane $\pi(N)$. By the property (3.13) all points of the lattice

$$L = \left\{P_1 + k_1\overrightarrow{P_1P_{r_1}} + k_2\overrightarrow{P_1P_{r_1}}\,;\ k_1, k_2 \text{ integers} \geqslant 0\right\} \tag{6.12}$$

are in Σ.

Let us now try to "cover" the entire plane $\pi(N)$ by translates of L with integer components. For convenience we shall refer to a translation

$$x_i = x_i' + n_i \qquad (i = 1,2,3), \tag{6.13}$$

as a *modular translation*, provided that the n_i are integers. Our plane

$$\pi(N): \sum_1^3 B_ix_i + B_4 = N \tag{6.14}$$

is changed by (6.13) into $\Sigma B_ix_i' + \Sigma B_in_i + B_4 = N$, and (6.14) is seen to be unchanged, or invariant, iff the n_i satisfy the equation

$$\sum_1^3 B_in_i = 0. \tag{6.15}$$

The modular translations (6.13) therefore correspond to those lattice points (n_1, n_2, n_3) which belong to the plane

$$\pi_0 : \sum_1^3 B_i x_i = 0. \tag{6.16}$$

We propose as a problem at the end of this chapter to show that the set

$$\Lambda = \left\{ (n_1, n_2, n_3);\ \sum_1^3 B_i n_i = 0,\ n_i \text{ integers} \right\} \tag{6.17}$$

is an (oblique) lattice of points of the plane π_0. Taking this for granted here, it is clear geometrically that we can find a point $Q \in \Lambda$ such as to have the vector equation

$$\overrightarrow{OQ} = \alpha_1 \overrightarrow{P_1 P_{r_1}} + \alpha_2 \overrightarrow{P_1 P_{r_2}} \quad \text{with} \quad \alpha_1 > 0, \alpha_2 > 0 \quad \text{(see Figure 16.1)}. \tag{6.18}$$

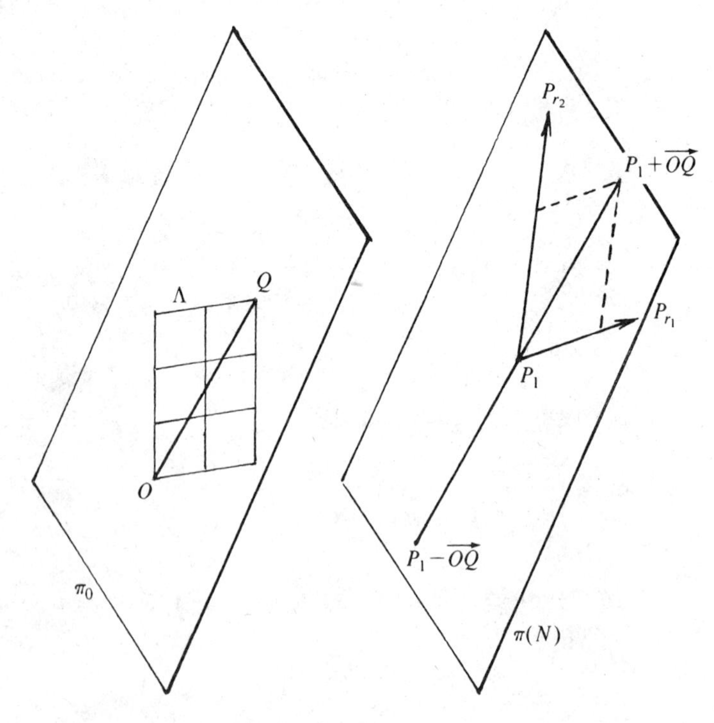

FIG. 16.1

But then if we perform on the set L, of (6.12), the sequence of translations

$$-q\overrightarrow{OQ} \qquad (q=1,2,\ldots), \tag{6.19}$$

then these translates of L will "cover" the plane $\pi(N)$ as far as we wish.

On the other hand, the sides of the triangle (6.11) are all $<4\delta$. Since L and all its translates by the vectors (6.19) are in Σ, it follows that the set $\Sigma \cap \pi(N)$ is dense in $\pi(N)$. This proves Lemma 4 and Lemma 3 establishes Theorem 6.

As in Theorem 4, our Theorem 6 generalizes to

THEOREM 7 (Kronecker). *Let the numbers* a_1, a_2, a_3 *satisfy the conditions* (4.3), (4.4), (4.5), (4.6) *of Theorem* 6. *Then the set* Σ_c *of points of* $\mathbb{R}^3$ *which are congruent* mod 1 *to points of the sequence*

$$A_n = (na_1 + c_1, na_2 + c_2, na_3 + c_3) \qquad (n=1,2,\ldots), \tag{6.20}$$

are contained in and are dense in the sequence of planes

$$\pi_c(N): \sum_1^3 B_i x_i = \sum_1^3 B_i c_i + N, \quad \textit{with } N \in \mathbb{Z}. \tag{6.21}$$

Proof: Clearly, the set Σ_c arises from Σ if we translate it by (c_1, c_2, c_3), so that we can write

$$\sum_c = \sum + (c_1, c_2, c_3).$$

But then Σ_c is contained in and is dense in the set of planes arising from the planes (4.12), and this is the set of planes (6.21).

Problems

1. The 2-dimensional analogue of Theorem 3 of §3 may be stated as follows: If the numbers $a_1, a_2, 1$, are linearly independent, then the points $P_n = ((na_1), (na_2))$, $n=1,2,\ldots,$ are dense in the square $\gamma_2 = \{0 \leqslant x_1 < 1, 0 \leqslant x_2 < 1\}$. Prove it.

Hint: Lettenmeyer's proof as presented here applies with suitable modifications and simplifications. See [**1**, §23.7].

2. This is the problem concerning the set Λ defined by (6.13). Let $\mathfrak{L}_3 = \{(n_1, n_2, n_3);\ n_i \in \mathbb{Z}\}$ be the lattice of $\mathbb{R}^3$. We consider the plane (6.15)

defined by

$$\pi_0 : \sum_1^3 B_i x_i = 0, \qquad B_i \text{ integers}, \quad \sum B_i^2 > 0,$$

and we are to show that the set

$$S = \mathfrak{L}_3 \cap \pi_0$$

forms a 2-dimensional lattice $\mathfrak{L}_2 \in \pi_0$. This means that there are in S two points u and v, spanning π_0, such that every $a \in S$ may be represented in the form

$$a = mu + nv, \quad \text{with integers } m \text{ and } n.$$

Hint: Establish the following statements:

(i) That the points of S form an additive group: *If* $a \in S$ *and* $b \in S$, *then also* $a - b \in S$, which implies that $0 \in S$ and $-a \in S$.

(ii) In view of (i), *the set* S *is invariant if we translate* S *by a vector* a, *where* $a \in S$.

(iii) Because $S \in \mathfrak{L}$, we have that $a \in S, b \in S, a \neq b$, *imply that* $|a - b| \geqslant 1$.

(iv) Using (iii), determine in S a point u such that

$$|u| = \min \quad |v| \quad \text{among all} \quad v \in S, v \neq 0.$$

Let λ denote the line joining 0 to u, and show that

$$\sigma = \{nu;\, n \in \mathbb{Z}\} = \lambda \cap S.$$

(v) Find in $S\sigma$ a point v which is nearest to the line λ. Using (ii) we see that in determining v it suffices to consider only those points of S, which are between the two normals to λ intersecting λ at 0 and u.

Using these remarks show that the representation $a = mu + nv$ holds.

References

[1] G. H. Hardy and E. M. Wright, *An Introduction to the Theory of Numbers*, 3rd ed., Oxford, 1954.

[2] F. Lettenmeyer, *Neuer Beweis des allgemeinen Kroneckerschen Approximationssatzes*, Proc. London Math. Soc. (2), 21 (1922) 306–314.

[3] I. Niven, *Irrational Numbers*, Carus Math. Monograph No. 11, Mathematical Association of America, 1956.

CHAPTER **17**

ON THE MOTIONS OF A BILLIARD BALL III: THE CONTINUOUS KRONECKER THEOREMS AND THE THEOREMS OF KÖNIG AND SZÜCS

1. Introduction. This is a continuation of Chapter 16 culminating in Theorems 11 and 12 of König and Szücs on the global properties of billiard ball motions. These are based on Kronecker's Theorems 5 and 7 of Chapter 16. However, before these can be used for this purpose, Theorems 5 and 7 have to be replaced by their "continuous" corollaries stated below as Theorems 9 and 10.

Oversimplifying the situation, we can state the contribution of König and Szücs in a single sentence: "The auxiliary function (x) of Figure 17.2(a) is being replaced by the auxiliary function $\langle x \rangle$ of Figure 17.2(b)."

However, because the linear Euler spline $\langle x \rangle$ is a *continuous function*, König and Szücs have thereby created an attractive new chapter of the theory of polyhedra.

2. The Kronecker Systems. Let

$$\pi: B_1x_1 + B_2x_2 + B_3x_3 = \gamma \tag{2.1}$$

be a given plane. The operation of "reducing the plane π mod 1" consists in replacing every point (x_1, x_2, x_3) of π, by its "least positive residue" $((x_1), (x_2), (x_3))$. This amounts to forming the set

$$K = \{((x_1), (x_2), (x_3)); (x_1, x_2, x_3) \in \pi\}. \tag{2.2}$$

This set assumes a simple and striking form, provided that we suitably restrict the nature of the coefficients B_i. We shall assume that

$$\textit{the } B_i \textit{ are integers}, \sum_1^3 B_i^2 > 0, \textit{ such that } (B_1, B_2, B_3) = 1. \tag{2.3}$$

Under these conditions we may state

LEMMA 5.* *Let the planes*

$$\pi(N): B_1x_1 + B_2x_2 + B_3x_3 = \gamma + N, \qquad N \in \mathbb{Z}, \tag{2.4}$$

be such that they intersect the cube $\gamma_3 = \{0 \leq x_i < 1, i = 1,2,3\}$, *hence*

$$\gamma_3 \cap \pi(N) \neq \emptyset, \tag{2.5}$$

and let $\{N\}$ *denote the set of values of* N *satisfying the nonvoid intersection property* (2.5). *We form these intersections*

$$F(N) = \gamma_3 \cap \pi(N) \quad \textit{for} \quad N \in \{N\}. \tag{2.6}$$

Then the set (2.2) *may be expressed as*

$$K = \bigcup_{N \in \{N\}} F(N). \tag{2.7}$$

We call K a *Kronecker system* (abbreviated to K-system), while the $F(N)$ are called its *facets*. We also say that K is *generated by the plane* π and write

$$K = K(\pi). \tag{2.8}$$

Proof: We dissect $\mathbb{R}^3$ into unit cubes

$$C(n_1, n_2, n_3): n_i \leq x_i < n_i + 1 \qquad (i = 1,2,3), \tag{2.9}$$

and choose the cubes such that

$$\pi(N) \cap C(n_1, n_2, n_3) \neq \emptyset. \tag{2.10}$$

If $x_i = (x_i) + [x_i]$, $(i = 1,2,3)$ is a point of the intersection (2.10), then $[x_i] = n_i$, and so $x_i = n_i + (x_i)$. To obtain points of K according to (2.2), we translate $C(n_1, n_2, n_3)$ into the cube $C(0,0,0) = \gamma_3$. This amounts to performing the translation by the vector $-(n_1, n_2, n_3)$, and the equation (2.1) is thereby changed into $\Sigma_1^3 B_i(x_i - n_i) = \gamma$; hence

$$\sum_1^3 B_i x_i = \gamma + \sum_1^3 B_i n_i. \tag{2.11}$$

---

*Because Chapters 16 and 17 are so closely related, we number lemmas and theorems consecutively in these two chapters. The symbol $\mathbb{Z}$ denotes the set of rational integers.

This equation is of the form (2.4). Conversely, if the number N in (2.4) is prescribed satisfying (2.5), then (2.3) shows that we can find the n_i such that $\Sigma B_i n_i = N$, and so (2.11) reduces to (2.4). Evidently, the set $\{N\}$ contains one or several *consecutive* elements.

EXAMPLE. Let us find the K-system generated by the plane

$$\pi: x_1 + x_2 + x_3 = \tfrac{1}{2}. \tag{2.12}$$

The condition (2.5) shows that $\{N\} = \{0,1,2\}$, and Figure 17.1 shows the 3 facets $F(0)$, $F(1)$, $F(2)$, and

$$K(\pi) = F(0) \cup F(1) \cup F(2). \tag{2.13}$$

Evident is the set equation

$$\pi = \cup \left(\pi \cap C(n_1, n_2, n_3)\right), \tag{2.14}$$

where we restrict the union to such cubes that intersect π. Just as two neighboring elements of this union have a common edge, so does the K-system (2.7) exhibit the following closure property: If we identify pointwise any two opposite facets of γ_3, then we can roam at will within K, passing from facet to facet, without ever encountering any boundary.

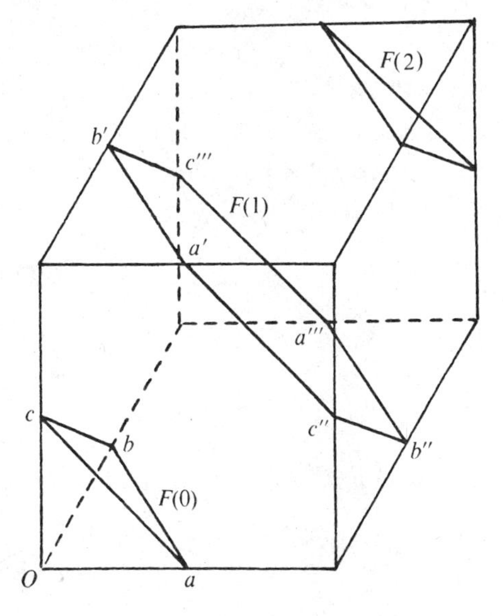

FIG. 17.1

In Figure 17.1 we identify the edge ab with $a'b'$; also bc with $b''c''$; ca with $c'''a'''$, a.s.f. For the plane (2.12) Figure 17.1 shows that the terms of the union (2.14) consist in regular triangles and regular hexagons.

3. A Reformulation of Theorem 7 of Chapter 16 in Terms of K-systems. We return to Theorem 7 of the previous chapter, which describes the location of the set Σ_c of points which are congruent mod 1 to the sequence of points

$$A_n = (na_1 + c_1, na_2 + c_2, na_3 + c_3) \qquad (n=1,2,\ldots). \tag{3.1}$$

We may as well replace this by the sequence of points

$$P_n = ((na_1 + c_1), (na_2 + c_2), (na_3 + c_3)) \qquad (n=1,2,\ldots), \tag{3.2}$$

and let

$$\Pi_c = \{P_n; n=1,2,\ldots\}. \tag{3.3}$$

Theorem 7 states that, writing $\gamma = \Sigma_1^3 B_i c_i$, the set Σ_c is contained in and is dense in the set of planes

$$\pi(N)\colon B_1x_1 + B_2x_2 + B_3x_3 = \gamma + N, \quad \text{where } N \in \mathbb{Z}. \tag{3.4}$$

Do the intersections of these planes with γ_3 form a K-system?

The answer is no, because the essential condition (2.3) is not necessarily satisfied. However, if we let

$$d = (B_1, B_2, B_3), \quad B_i' = \frac{B_i}{d}, \quad \gamma' = \frac{\gamma}{d}, \quad N = N'd + \nu \quad (0 \leq \nu \leq d-1), \tag{3.5}$$

where N' is the quotient of N by d, and ν is the remainder, then we may replace the equations (3.4) by

$$\Pi_\nu(N')\colon B_1'x_1 + B_2'x_2 + B_3'x_3 = \gamma' + \frac{\nu}{d} + N' \qquad (N' \in \mathbb{Z}; 0 \leq \nu \leq d-1). \tag{3.6}$$

We have therefore derived from Theorem 7 the following

THEOREM 8. *We assume that the numbers $a_1, a_2, a_3, 1$ are connected by the unique equation*

$$B_1a_1 + B_2a_2 + B_3a_3 + B_4 = 0 \qquad (B_1, B_2, B_3, B_4) = 1 \tag{3.7}$$

with integer coefficients. Then the sequence of points (3.2) *is contained in and is dense in the facets of d K-systems* K_ν $(\nu=0,\ldots,d-1)$, *where* $K_\nu=K(\pi_\nu)$ *is generated by the plane*

$$\pi_\nu:\ B_1'x_1+B_2'x_2+B_3'x_3=\frac{1}{d}\sum_1^3 B_ic_i+\frac{\nu}{d}\qquad(\nu=0,\ldots,d-1).\quad(3.8)$$

A proof follows from $(B_1', B_2', B_3')=1$ and the set equation

$$\Pi_c=\gamma_3\cap\Sigma_c.$$

4. The Continuous Kronecker Theorems. So far we have dealt with the distribution mod 1 of equidistant sequences (3.1). Having in mind applications to K-S polygons, we now derive as corollaries of Theorems 5 and 8 the location mod 1 of rectilinear motions.

THEOREM 9. *Let the numbers*

$$\lambda_1,\lambda_2,\lambda_3 \textit{ be linearly independent.}\tag{4.1}$$

Then the path Λ *of the piecewise rectilinear motion*

$$\Lambda(t)=((\lambda_1t+c_1),(\lambda_2t+c_2),(\lambda_3t+c_3))\qquad(t\geqslant 0),\tag{4.2}$$

is dense in the cube γ_3.

Proof: We consider the linear form $l(r_i)=\lambda_1r_1+\lambda_2r_2+\lambda_3r_3$, and the set

$$S=\{l(r_1,r_2,r_3);\ \text{all } r_i \text{ assume rational values}\}.\tag{4.3}$$

This being a countable set, its complement $M=\mathbb{R}\setminus S$ is nonvoid. If μ is positive and $\mu\in M$, then the four reals $\lambda_1,\lambda_2,\lambda_3,\mu$ are linearly independent, and this implies that

$$\frac{\lambda_1}{\mu},\quad\frac{\lambda_2}{\mu},\quad\frac{\lambda_3}{\mu},\quad 1\quad\text{are linearly independent.}\tag{4.4}$$

To the numbers $a_i=\lambda_i/\mu$ we may apply Theorem 5 to conclude that the points

$$P_n(\mu)=\left(\left(n\frac{\lambda_1}{\mu}+c_1\right),\left(n\frac{\lambda_2}{\mu}+c_2\right),\left(n\frac{\lambda_3}{\mu}+c_3\right)\right)\ \text{are dense in } \gamma_3.\tag{4.5}$$

This being a subsequence of points of the path Λ of (4.2), it follows that Λ is dense in γ_3.

Our next theorem will be a consequence of Theorem 8. We drop the assumption (4.1) and assume instead that there is a linear equation

$$B_1\lambda_1 + B_2\lambda_2 + B_3\lambda_3 = 0, \qquad B_i \textit{ are integers}, \quad (B_1, B_2, B_3) = 1, \tag{4.6}$$

and that

there is no second equation of the form (4.6) *independent of* (4.6). (4.7)

Again we select a positive μ in the complement of the set (4.3), so that (4.4) holds. By Theorem 8 we conclude that the sequence (4.5) is contained densely in the K-system generated by the plane

$$\pi: B_1x_1 + B_2x_2 + B_3x_3 = \sum_1^3 B_ic_i \qquad (B_1, B_2, B_3) = 1. \tag{4.8}$$

Notice that by (4.6) we have $d = 1$, and for this reason we obtain only one K-system. Again, the points (4.5) belong to Λ and we have established

THEOREM 10. *If* $\lambda_1, \lambda_2, \lambda_3$ *are connected by the unique linear equation* (4.6), *then the path* Λ *of the piecewise rectilinear motion*

$$\Lambda(t) = ((\lambda_1 t + c_1), (\lambda_2 t + c_2), (\lambda_3 t + c_3)) \qquad (t \geqslant 0) \tag{4.9}$$

is contained in and is dense in the K-system generated by the plane (4.8).

5. On K-systems and K-S Polyhedra Generated by the Same Plane π. Let

$$\pi: B_1x_1 + B_2x_2 + B_3x_3 = \gamma, \quad B_i \text{ are integers}, \tag{5.1}$$

such that

$$\sum_1^3 B_i^2 > 0 \qquad (B_1, B_2, B_3) = 1, \tag{5.2}$$

be a prescribed plane. In §2 we have defined the K-system

$$K = K(\pi) \tag{5.3}$$

generated by the plane π.

In §4 of Chapter 15 we have defined the K-S polyhedron

$$\Pi^{\cdot} = \Pi^2(\pi) \tag{5.4}$$

obtained by successive reflections of the plane π in the facets of γ_3. (In Chapter 15 the plane π was denoted by L_2.)

We saw in (2.2) how $K(\pi)$ can be defined in terms of the auxiliary function

$$(x) = x - [x], \tag{5.5}$$

having the graph of Figure 17.2(a). However, also $\Pi^2(\pi)$ may be described in terms of the linear Euler spline $\langle x \rangle$ defined by

$$\langle x \rangle = |x| \quad \text{if} \quad -1 \leqslant x \leqslant 1 \quad \text{and} \quad \langle x+2 \rangle = \langle x \rangle \quad \text{for all} \quad x, \tag{5.6}$$

and graphed in Figure 17.2(b). In terms of these functions, the definitions of $K(\pi)$ and $\Pi^2(\pi)$ are

$$K(\pi) = \{((x_1),(x_2),(x_3)); \quad (x_1,x_2,x_3) \in \pi\} \tag{5.7}$$

and

$$\Pi^2(\pi) = \{(\langle x_1 \rangle, \langle x_2 \rangle, \langle x_3 \rangle); \quad (x_1,x_2,x_3) \in \pi\}. \tag{5.8}$$

If $\pi = L_2$ is represented parametrically by the equations (4.5) of Chapter 15, then $\Pi^2(\pi)$ is represented by the equations (4.9) of Chapter 15, and $K(\pi)$ may be similarly represented using the function $(\cdot)$.

A more elaborate description of the relation between $K(\pi)$ and $\Pi^2(\pi)$, to be used below in §7, is as follows. We consider again the unit cubes (2.9), in particular those such that

$$\pi \cap C(n_1, n_2, n_3) \neq 0. \tag{5.9}$$

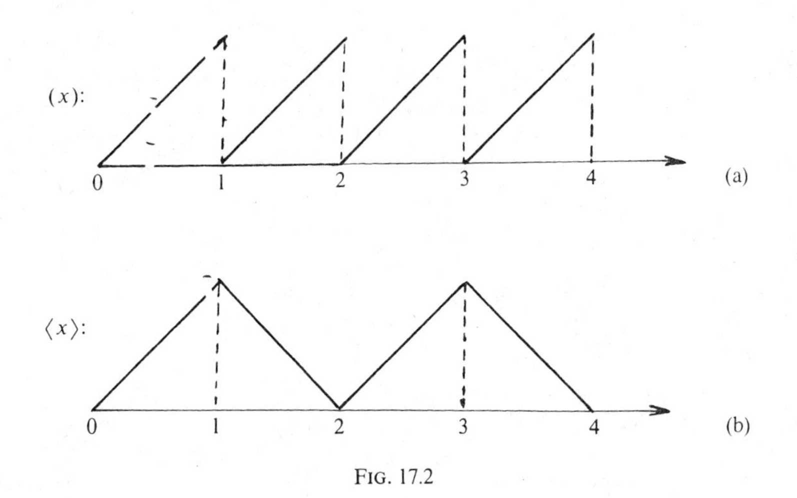

FIG. 17.2

By the translation $C(n_1, n_2, n_3) \to C(0,0,0) = \gamma_3$ the intersection (5.9) becomes a facet F of $K(\pi)$. *How do we get a facet f of $\Pi^2(\pi)$?*

A comparison of the graphs of Figure 17.2 shows that

$$\langle x \rangle = \begin{cases} (x) & \text{if } [x] \text{ is even,} \\ 1-(x) & \text{if } [x] \text{ is odd.} \end{cases} \tag{5.10}$$

In geometric terms we see from (5.10), and the representations (5.8) and (5.7), that a facet F of $K(\pi)$ is changed into a facet f of $\Pi^2(\pi)$ by an appropriate reflection of γ_3 into itself. For instance, the reflection of $\gamma_3 \to \gamma_3$ is

$$\begin{cases} \text{the identity} & (x_1, x_2, x_3) \to (x_1, x_2, x_3) \\ & \text{if } n_1, n_2, n_3 \text{ are all even,} \\ \text{the reflection} & (x_1, x_2, x_3) \to (1-x_1, x_2, x_3) \\ & \text{if } n_1 \text{ is odd, } n_2 \text{ and } n_3 \text{ are even,} \\ \text{the reflection} & (x_1, x_2, x_3) \to (1-x_1, 1-x_2, x_3) \\ & \text{if } n_1 \text{ and } n_2 \text{ are odd, } n_3 \text{ is even,} \end{cases} \tag{5.11}$$

and so forth. The rule seems obvious: *We obtain f from F by leaving x_i unchanged, if $[x_i]$ is even, while changing x_i into $1-x_i$, if $[x_i]$ is odd.*

This makes it plain that the facets of $\Pi^2(\pi)$ are congruent to facets of $K(\pi)$ and conversely. However, the structure and aspect of $K(\pi)$ and $\Pi^2(\pi)$ are very different. This is nicely shown by

AN EXAMPLE. The $K(\pi)$ generated by the plane

$$\pi: x_1 + x_2 + x_3 = \tfrac{1}{2}$$

is seen in Figure 17.1 to have three faces $F(0)$, $F(1)$, $F(2)$. The K-S polyhedron $\Pi^2(\pi)$ is shown in Figure 15.10 of Problem 3 of Chapter 15, and is seen to have 12 facets, of which 8 are regular triangles and 4 are regular hexagons.

6. The Two Theorems of König and Szücs. We are finally able to establish the König-Szücs results already stated in §4 of Chapter 15 in the form of the dichotomy (i),(ii). They were there stated, without their assumptions, to motivate the introduction of the K-S polyhedra.

The theorems are as follows:

THEOREM 11 (König and Szücs). *Let the numbers*

$$\lambda_1, \lambda_2, \lambda_3 \textit{ be linearly independent.} \tag{6.1}$$

Then the K-S *polygon*

$$\Pi:\ \begin{aligned} x_1 &= \langle \lambda_1 t + c_1 \rangle \\ x_2 &= \langle \lambda_2 t + c_2 \rangle \\ x_3 &= \langle \lambda_3 t + c_3 \rangle \qquad (t \geqslant 0) \end{aligned} \tag{6.2}$$

is dense in the cube γ_3.

THEOREM 12 (König and Szücs). *Let* $\lambda_1, \lambda_2, \lambda_3$ *be connected by the equation*

$$B_1\lambda_1 + B_2\lambda_2 + B_3\lambda_3 = 0, \quad B_i \textit{ integers}, \quad (B_1, B_2, B_3) = 1, \tag{6.3}$$

such that

there is no second equation of the form (6.3) *independent of* (6.3). (6.4)

Then the K-S *polygon* (6.2) *is contained in and is dense in the* K-S *polyhedron* $\Pi^2(\pi)$ *generated by the plane*

$$\pi: B_1x_1 + B_2x_2 + B_3x_3 = \sum_1^3 B_ic_i. \tag{6.5}$$

Out of order we begin with Theorem 12 because, in the light of our remarks of §5, it is an immediate consequence of Theorem 10. We have found there that the facets of $\Pi^2(\pi)$ arise from those of $K(\pi)$ by appropriate reflections of γ_3 into itself of the type described by (5.11). However, also the edges of the polygon Π, of (6.2), arise from those of the polygon Λ described by $\Lambda(t)$, of (4.9), by the same reflections of the type (5.11). By Theorem 10 we have $\Lambda \subset K(\pi)$ and Λ is dense in $K(\pi)$. Therefore Π is densely contained in $\Pi^2(\pi)$, proving Theorem 12.

7. A Proof of Theorem 11. The transition from Kronecker's theorems to his continuous theorems and to the König-Szücs Theorem 12 has been so smooth that it seems perhaps fitting that there be a little struggle to derive their Theorem 11.

The polygon Λ described by

$$\Lambda(t) = ((\lambda_1 t + c_1), (\lambda_2 t + c_2), (\lambda_3 t + c_3)) \qquad (t \geqslant 0), \tag{7.1}$$

has all its sides parallel to the original half-line

$$\lambda: x_1 = \lambda_1 t + c_1, \quad x_2 = \lambda_2 t + c_2, \quad x_3 = \lambda_3 t + c_3 \qquad (t \geqslant 0), \tag{7.2}$$

and by Theorem 9 Λ is dense in γ_3. The half-line λ is dissected into disjoint intervals according to the set equation

$$\lambda = \bigcup_{n_i} (\lambda \cap C(n_1, n_2, n_3)), \tag{7.3}$$

where we use only cubes such that

$$I(n_1, n_2, n_3) = \lambda \cap C(n_1, n_2, n_3) \neq \varnothing. \tag{7.4}$$

The interval $I(n_1, n_2, n_3)$ is also characterized by the simultaneous inequalities

$$n_1 \leqslant \lambda_1 t + c_1 < n_1 + 1, \quad n_2 \leqslant \lambda_2 t + c_2 < n_2 + 1,$$

$$n_3 \leqslant \lambda_3 t + c_3 < n_3 + 1. \tag{7.5}$$

The edges of Λ arise from the intervals $I(n_1, n_2, n_3)$ by the translation

$$C(n_1, n_2, n_3) \to C(0,0,0) = \gamma_3. \tag{7.6}$$

However, for the polygon Π described by

$$\Pi(t) = (\langle \lambda_1 t + c_1 \rangle, \langle \lambda_2 t + c_2 \rangle, \langle \lambda_3 t + c_3 \rangle) \qquad (t \geqslant 0), \tag{7.7}$$

the situation is different and depends on the parity of the n_i. A comparison of the graphs of Figure 17.2 shows that

$$\langle x \rangle = \begin{cases} (x) & \text{if} \quad [x] \text{ is even,} \\ 1-(x) & \text{if} \quad [x] \text{ is odd.} \end{cases} \tag{7.8}$$

Therefore, if

$$n_1, n_2, n_3 \text{ are all even,} \tag{7.9}$$

then the corresponding sides of Λ and Π are identical, as they both arise from $I(n_1, n_2, n_3)$ by the same translation (7.6) in view of (7.5) and (7.8). This is not the case if some n_i is odd.

I claim: *Already the edges of* Π *corresponding to the case* (7.9), *and therefore identical with the edges of* Λ, *are dense in* γ_3.

It is evidently sufficient to prove:

$$\textit{The edges of } \Lambda \textit{ for the case } (7.9), \textit{ are already dense in } \gamma_3. \tag{7.10}$$

Proof: Let P be a point of γ_3. By Theorem 9 there is a point $\Lambda(t)$ arbitrarily near to P. Assume that the inequalities (7.5) hold. If all n_i are even, there is nothing to prove. If not, we will now perturb t to the value $t+\tau$ such that

(i) The point $\Lambda(t+\tau)$ is also arbitrarily near to P.
(ii) The inequalities (7.5) for the new point are

$$m_1 < \lambda_1(t+\tau)+c_1 < m_1+1, \quad m_2 < \lambda_2(t+\tau)+c_2 < m_2+1,$$

$$m_3 < \lambda_3(t+\tau)+c_3 < m_3+1, \tag{7.11}$$

where

$$m_1, m_2, m_3 \text{ are all even.} \tag{7.12}$$

It suffices to discuss the case when in (7.5)

$$n_1 \textit{ is odd, while } n_2 \textit{ and } n_3 \textit{ are even.} \tag{7.13}$$

How do we select τ *so that in* (7.11) m_1, m_2, m_3 *are all even*? This is easily done as follows. By a slight change of t, if necessary, we may replace (7.5) by the strict inequalities

$$n_1 < \lambda_1 t + c_1 < n_1+1, \quad n_2 < \lambda_2 t + c_2 < n_2+1,$$

$$n_3 < \lambda_3 t + c_3 < n_3+1. \tag{7.14}$$

The numbers $\frac{1}{2}\lambda_1, \frac{1}{2}\lambda_2, \frac{1}{2}\lambda_3$ are also linearly independent. In Theorem 9 we select $c_1 = c_2 = c_3 = 0$, and can determine a value τ, such that the point $((\frac{1}{2}\lambda_1\tau),(\frac{1}{2}\lambda_2\tau),(\frac{1}{2}\lambda_3\tau))$ is as close to the point $(\frac{1}{2},0,0)$ of γ_3, as we wish. This means that there are integers N_i such that

$$\tfrac{1}{2}\lambda_1\tau = N_1 + \tfrac{1}{2} + \tfrac{1}{2}\epsilon_1,$$

$$\tfrac{1}{2}\lambda_2\tau = N_2 \qquad + \tfrac{1}{2}\epsilon_2,$$

$$\tfrac{1}{2}\lambda_3\tau = N_3 \qquad + \tfrac{1}{2}\epsilon_3,$$

where the ϵ_i are as small as we wish, and therefore

$$\begin{aligned} \lambda_1\tau &= 2N_1 + 1 + \epsilon_1, \\ \lambda_2\tau &= 2N_2 \qquad + \epsilon_2, \\ \lambda_3\tau &= 2N_3 \qquad + \epsilon_3. \end{aligned} \tag{7.15}$$

Adding these to the corresponding inequalities (7.14), and assuming the ϵ_i

sufficiently small, we obtain

$$\begin{aligned} n_1+(2N_1+1) &< \lambda_1(t+\tau)+c_1 < n_1+(2N_1+1)+1 \\ n_2+2N_2 &< \lambda_2(t+\tau)+c_2 < n_2+2N_2+1 \\ n_3+2N_3 &< \lambda_3(t+\tau)+c_3 < n_3+2N_3+1, \end{aligned}$$

and we see from (7.13) that the inequalities (7.11), *for the point* $\Lambda(t+\tau)$, satisfy the condition (7.12). Thus we have satisfied the condition (ii). However, also the condition (i) is satisfied in view of (7.15), and the fact that the function (x) has the period 1.

Obvious modification will take care of the other possibilities. For instance, if we replace (7.13) by the assumption that n_1 *and* n_2 *are odd, while* n_3 *is even,* then we select τ, by Theorem 9, such that the point $((\frac{1}{2}\lambda_1\tau),(\frac{1}{2}\lambda_2\tau),(\frac{1}{2}\lambda_3\tau))$ *is close to the point* $(\frac{1}{2},\frac{1}{2},0)$ *of* γ_3.

Problems

1. The following is an instructive numerical example of Theorem 8 of §3. It becomes also enjoyable if we use a hand-held calculator with a memory place. In Theorem 8 we choose

(1) $a_1=\frac{1}{8}\sqrt{2}=.17678$, $a_2=\frac{1}{16}\sqrt{3}=.10825$, $a_3=\frac{1}{24}(4-\sqrt{2}-\sqrt{3})=.03557$.

(i) Verify that $2a_1+4a_2+6a_3-1=0$ is the unique linear equation (3.7) between the a_i. Here $d=(B_1,B_2,B_3)=2$ and therefore two K-systems should appear.

(ii) Using the 5-place values (1) construct a table of values of the multiples na_1, na_2, na_3, for $n=1,2,\ldots,30$.

Derive from it a table of values of

$$(na_1)+2(na_2)+3(na_3) \quad \text{for } n=1,\ldots,30$$

and verify that the points $P_n=((na_1),(na_2),(na_3))$ are located on the two K-systems

$$K_0\colon x_1+2x_2+3x_3=N \quad \text{for } N=1,2,3,4,5$$

$$K_1\colon x_1+2x_2+3x_3=\tfrac{1}{2}+N \quad \text{for } N=0,1,2,3,4,5.$$

Notice that the 30 points P_n are slightly off these planes due to rounding errors. Thus for $n=27$ we find that $(na_1)+2(na_2)+3(na_3)=5.49973$ instead of 5.5.

2. In Theorem 12 of §6 let

$$\lambda_1=\tfrac{1}{4}\sqrt{2}, \quad \lambda_2=\tfrac{1}{4}\sqrt{3}, \quad \lambda_1=-\tfrac{1}{4}\left(\sqrt{2}+\sqrt{3}\right)$$

and

$$c_1 = 1, \quad c_2 = 0, \quad c_3 = 0.$$

Show that $\lambda_1 + \lambda_2 + \lambda_3 = 0$ is the unique equation (6.3), and that the K-S polygon

$$x_1 = \langle \lambda_1 t + 1 \rangle, \quad x_2 = \langle \lambda_2 t \rangle, \quad x_3 = \langle \lambda_3 t \rangle$$

is contained in and is dense in Kepler's regular tetrahedron $T = ABCD$ of Figure 18.1 of the next chapter.

References

[1] G. H. Hardy and E. M. Wright, *An Introduction to the Theory of Numbers*, 3rd ed., Oxford, 1954.

[2] D. König and A. Szücs, *Mouvement d'un point abandonné à l'intérieur d'un cube*, Rend. Circ. Mat. Palermo, 36 (1913) 79–90.

CHAPTER **18**

ON THE MOTIONS OF A BILLIARD BALL IV: AN EXTREMUM PROPERTY OF KEPLER'S STELLA OCTANGULA

1. Introduction. This fourth and last chapter of this series may be read independently. We recall the description of the Stella Octangula (S.O.) already introduced in Problem 3(i) at the end of Chapter 15 and shown in Figure 15.9 and also in Figure 18.1 below. It is composed of the surfaces of the two tetrahedra

$$T = ABCD, \quad T' = A'B'C'D' \tag{1.1}$$

inscribed in the unit cube γ_3; hence

$$\text{S.O.} = T \cup T'. \tag{1.2}$$

Let $c = (\frac{1}{2}, \frac{1}{2}, \frac{1}{2})$ be the center of γ_3, and let δ denote the center of the regular triangle ABC. Since

$$A = (1,0,0), \quad B = (0,1,0), \quad C = (0,0,1),$$

their centroid is $\delta = (\frac{1}{3}, \frac{1}{3}, \frac{1}{3})$, and so

$$\overrightarrow{c\delta} = (-\tfrac{1}{6}, -\tfrac{1}{6}, -\tfrac{1}{6}).$$

By the symmetries of γ_3 about its center c we conclude the following. If we write $x = (x_1, x_2, x_3)$ and $\|x\|_\infty = \max(|x_1|,|x_2|,|x_3|)$, then the open cube

$$C^*\colon \|x - c\|_\infty < \tfrac{1}{6} \tag{1.3}$$

is interior to both T and T', while its eight vertices are the centers of the eight faces of the tetrahedra.

Now we recall the definition of the *König-Szücs polyhedra* as defined in §4 of Chapter 15 (see [2]). We start from a plane

$$L_2\colon A_1x_1 + A_2x_2 + A_3x_3 = C \tag{1.4}$$

intersecting the cube γ_3, and reflect this plane, and also its reflections,

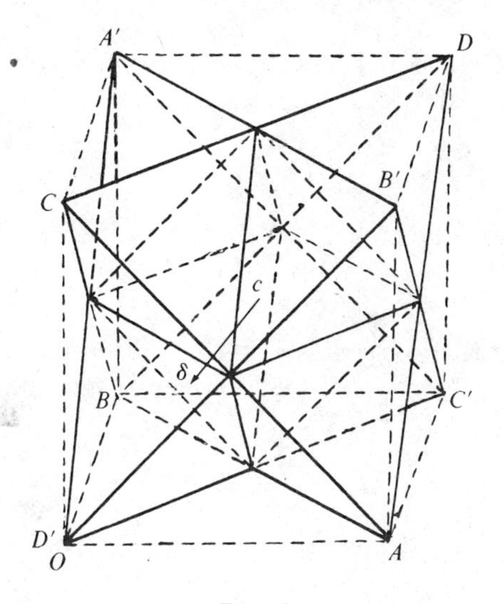

FIG. 18.1

whenever these strike any of the six facets of γ_3. It helps to visualize the situation if we think of L_2 as carrying a flat pencil of light-rays, starting from one of its points, which are reflected back into γ_3 whenever the rays strike any of its six mirrorlike facets. The way the individual rays are reflected is shown in Figure 15.6(a),(b),(c). Let

$$\Pi = \Pi(L_2) \tag{1.5}$$

denote the polyhedron obtained by all these reflections of L_2. The polyhedron (1.5) is by definition a *König-Szücs* (K-S) *polyhedron*. It is a finite or infinite polyhedron and $\Pi \subset \gamma_3$.

The simplest possible nontrivial K-S polyhedra are the tetrahedra (1.1). Indeed, we see from Figure 18.1 that

$$\Pi = T \quad \text{if} \quad L_2 \text{ is the plane } ABC \text{ (or any facet of } T\text{)}, \tag{1.6}$$

and

$$\Pi = T' \quad \text{if} \quad L_2 \text{ is the plane } A'B'C' \text{ (or any facet of } T'\text{)}. \tag{1.7}$$

An important new concept is this: *We say that the plane* (1.4) *is in general position, provided that*

$$A_1 \neq 0, \quad A_2 \neq 0, \quad A_3 \neq 0, \tag{1.8}$$

when we also say that $\Pi = \Pi(L_2)$ *is in general position* (or G.P.).

Evidently T and T' are K-S polyhedra in G.P. Moreover, neither T, nor T', ever penetrates into the open cube (1.3); hence

$$T \cap C^* = \varnothing, \qquad T' \cap C^* = \varnothing. \tag{1.9}$$

The extremum property of the S.O., which is the main result of this chapter, is the following:

THEOREM 1. *If the* K-S *polyhedron* $\Pi = \Pi(L_2)$ *is in general position, and* L_2 *is different from the eight facets of* (1.2), *then* Π *must penetrate into the cube* (1.3), *hence*

$$\Pi \cap C^* \neq \varnothing. \tag{1.10}$$

Here are two remarks which we mention without further discussion.

1. *Theorem* 1 *remains correct if in its statement we replace the cube* C^* *by its circumscribed open sphere*

$$S^*: \left(x_1 - \tfrac{1}{2}\right)^2 + \left(x_2 - \tfrac{1}{2}\right)^2 + \left(x_3 - \tfrac{1}{2}\right)^2 < \tfrac{1}{12}. \tag{1.11}$$

2. A result like Theorem 1 has higher- and also lower-dimensional analogues in γ_n and γ_2, respectively. For its analogue in γ_n see the forthcoming paper [5]. For its analogue in γ_2 we refer to Figure 15.2(c). The analogue of the S.O. $= T \cup T'$ is the square π of Figure 15.2(c) whose vertices are the midpoints of the sides of γ_2. *The square* π *does not penetrate into the open square*

$$s^*: \|x - c\|_\infty < \tfrac{1}{4}, \tag{1.12}$$

while any other K-S *polygon in general position* (i.e., with no sides parallel to the axes) *must intersect* s^*. This is the two-dimensional analogue of Theorem 1. See Problem 2 at the end of this chapter.

The main idea in the proof of Theorem 1 is to transfer the problem from the space R^3 to the plane R^2 by means of certain plane geometric objects which are called *monochromes* and 3-*chromos*.

2. A Few Definitions and Examples. We begin by replacing the equation (1.4) by a parametric representation

$$L_2: \begin{aligned} x_1 &= \lambda^1_1 u_1 + \lambda^2_1 u_2 + a_1, \\ x_2 &= \lambda^1_2 u_1 + \lambda^2_2 u_2 + a_2, \\ x_3 &= \lambda^1_3 u_1 + \lambda^2_3 u_2 + a_3, \qquad -\infty < u_1, u_2 < \infty, \end{aligned} \tag{2.1}$$

where $a=(a_1, a_2, a_3)$ is a point interior to γ_3. The conditions (1.8) for L_2 to be in general position are now replaced by the equivalent conditions that

$$\begin{vmatrix} \lambda_1^1 & \lambda_1^2 \\ \lambda_2^1 & \lambda_2^2 \end{vmatrix} \neq 0, \quad \begin{vmatrix} \lambda_1^1 & \lambda_1^2 \\ \lambda_3^1 & \lambda_3^2 \end{vmatrix} \neq 0, \quad \begin{vmatrix} \lambda_2^1 & \lambda_2^2 \\ \lambda_3^1 & \lambda_3^2 \end{vmatrix} \neq 0. \tag{2.2}$$

The advantage of the new representation (2.1) is that the reflections of the plane signal (2.1) are automatically achieved by using the *linear Euler spline* $\langle x \rangle$ defined by

$$\langle x \rangle = |x| \quad \text{if} \quad -1 \leqslant x \leqslant 1, \langle x+2 \rangle = \langle x \rangle \quad \text{for all real } x. \tag{2.3}$$

Its graph is shown in Figure 15.4 (also in Figure 18.7(a)). There it is also explained, in §2 of Chapter 15, why the K-S polyhedron Π, obtained by reflecting (2.1), is parametrically represented by the equations

$$\Pi: \begin{aligned} x_1 &= \langle \lambda_1^1 u_1 + \lambda_1^2 u_2 + a_1 \rangle, \\ x_2 &= \langle \lambda_2^1 u_1 + \lambda_2^2 u_2 + a_2 \rangle, \\ x_3 &= \langle \lambda_3^1 u_1 + \lambda_3^2 u_2 + a_3 \rangle, \qquad (-\infty < u_1, u_2 < \infty). \end{aligned} \tag{2.4}$$

A few examples are called for.

(i) By (1.6) we know that we get

$$\Pi = T, \tag{2.5}$$

if we choose for L_2 the plane ABC (Figure 18.1) having the equation $x_1 + x_2 + x_3 = 1$. Its parametric equations are

$$x_1 = u_1, \quad x_2 = u_2, \quad x_3 = 1 - u_1 - u_2,$$

and the equations (2.4) become

$$x_1 = \langle u_1 \rangle, \quad x_2 = \langle u_2 \rangle, \quad x_3 = \langle 1 - u_1 - u_2 \rangle.$$

From the properties of the Euler spline we have the identities

$$\langle 1 - u_1 - u_2 \rangle = \langle -1 - u_1 - u_2 \rangle = \langle 1 + u_1 + u_2 \rangle,$$

which show that

$$T: \begin{aligned} x_1 &= \langle u_1 \rangle, \\ x_2 &= \langle u_2 \rangle, \\ x_3 &= \langle u_1 + u_2 + 1 \rangle. \end{aligned} \tag{2.6}$$

(ii) Similarly, by (1.7), if we choose for L_2 the plane $A'B'C'$ having the equation $x_1+x_2+x_3=2$, we obtain for T' the representation

$$T': \begin{aligned} x_1 &= \langle u_1 \rangle, \\ x_2 &= \langle u_2 \rangle, \\ x_3 &= \langle u_1+u_2 \rangle. \end{aligned} \tag{2.7}$$

We return to our general discussion and choose ρ such that $0<\rho<\frac{1}{2}$. We also write $x=(x_1,x_2,x_3)$, $c=(\frac{1}{2},\frac{1}{2},\frac{1}{2})$, and consider the open cube

$$C_\rho: \|x-c\|_\infty < \rho. \tag{2.8}$$

DEFINITION 1. *We say that the* K-S *polyhedron* Π *is* ρ*-admissible, provided that* Π *is in general position, and such that* Π *never penetrates into the open cube* (2.8); *hence*

$$\Pi \cap C_\rho = \varnothing. \tag{2.9}$$

We are now trying to make the cube C_ρ as large as possible. With this in mind we define

$$\rho^* = \sup\{\rho;\ \text{among all } \rho \text{ having } \rho\text{-admissible K-S polyhedra } \Pi \text{ in general position}\}. \tag{2.10}$$

In terms of this quantity our Theorem 1 *is equivalent to the equation*

$$\rho^* = \tfrac{1}{6}. \tag{2.11}$$

Actually (2.11) only shows that there is no Π in general position which is ρ-admissible for any $\rho>\frac{1}{6}$. *We must also show, however, that* T *and* T' *are the only* K-S *polyhedra which are in general position and are* $\frac{1}{6}$*-admissible.*

To conclude this section we wish to point out the important role played by the condition "in general position" in the definition (2.10) of ρ^*. For, if we drop the requirement of G.P., then the value of the supremum (2.10) will actually exceed the value $\rho^*=\frac{1}{6}$.

This is shown if we choose the plane

$$L_2': x_1+x_2=\tfrac{1}{2}. \tag{2.12}$$

This is the plane ABB_1A_1 of Figure 18.2(a). The corresponding K-S polyhedron $\Pi'=\Pi(L_2')$ is seen to be composed of the four rectangles

$$\Pi' = ABB_1A_1 \cup A_1B_1B_2A_2 \cup A_2B_2B_3A_3 \cup A_3B_3BA.$$

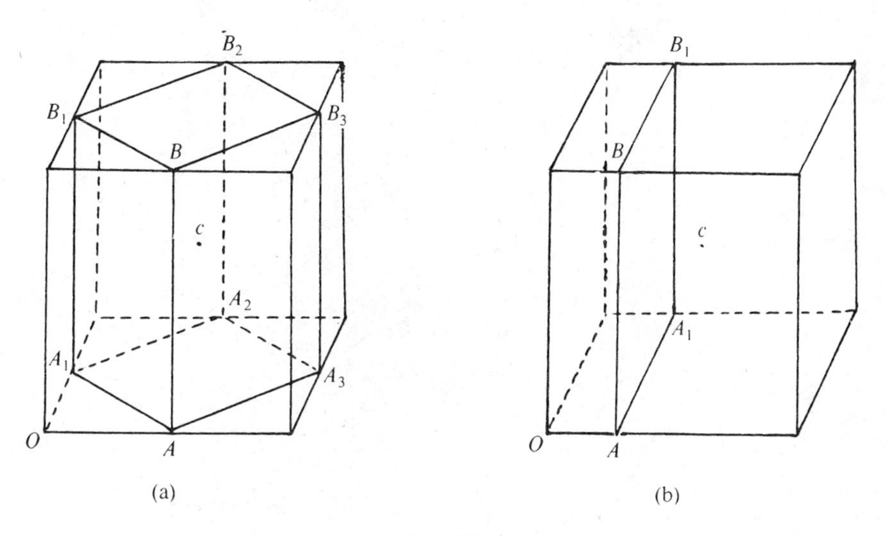

FIG. 18.2

A glance at Figure 18.2(a) shows that Π' is ρ'-admissible with $\rho' = \frac{1}{4}$, and $\rho' > \rho^* = \frac{1}{6}$. What is wrong? The answer: Π' is not in general position, because a comparison of (2.12) with (1.4) shows that $A_3 = 0$.

We obtain an even worse violation in Figure 18.2(b). Here we start from the plane L_2'': $x_1 = \epsilon$, with ϵ positive and small. This produces the reflected Π'' composed of the single rectangle ABB_1A_1. Here Π'' is clearly ρ''-admissible, with $\rho'' = \frac{1}{2} - \epsilon$, which is as close to $\frac{1}{2}$ as we wish; hence again $\rho'' > \rho^* = \frac{1}{6}$.

3. Monochromes and Chromos. A perfect example of what we wish to call a *monochrome* (shortened to MC) is an awning of the kind used to provide shade to storefronts. A monochrome is a sequence of parallel, congruent and equidistant strips in the plane R^2 of a rectangular coordinate system (u_1, u_2); see Figure 18.3(a). This is a periodic structure, its *period* p being the distance between similar boundary lines of two conservative strips. Let w be the *width* of a strip. We like to think of the strips as carrying a certain color γ, and this explains the name monochrome (derived from $\chi\rho\hat{\omega}\mu\alpha$ = color). The system of lines dividing each strip into equal halves we call the *central lines* of the MC. The most important quantity attached to an MC is the ratio

$$\delta = \frac{w}{p}, \tag{3.1}$$

called *the density* of the MC, and we denote the MC by the symbol

$$M(\delta).$$

Clearly, the density δ remains unchanged if we subject $M(\delta)$ to an affine transformation. In this sense the invariant δ characterizes the MC.

Figure 18.3(a) shows $M(\frac{1}{2})$, because $p = 2w$. Figure 18.3(b) exhibits *three* monochromes

$$M_1(\tfrac{1}{3}),\quad M_2(\tfrac{1}{3}),\quad M_3(\tfrac{1}{3}), \tag{3.2}$$

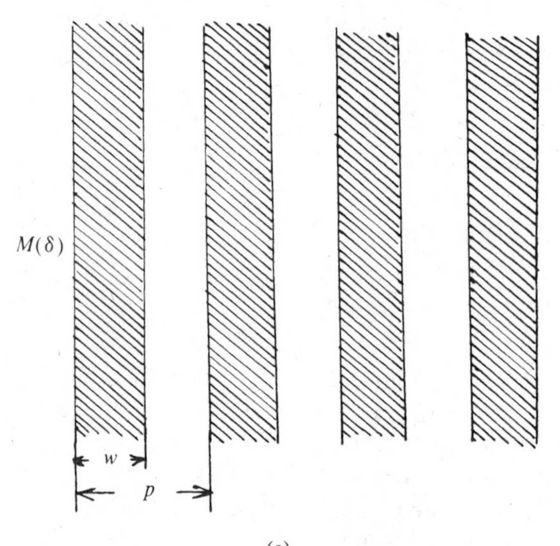

(a)

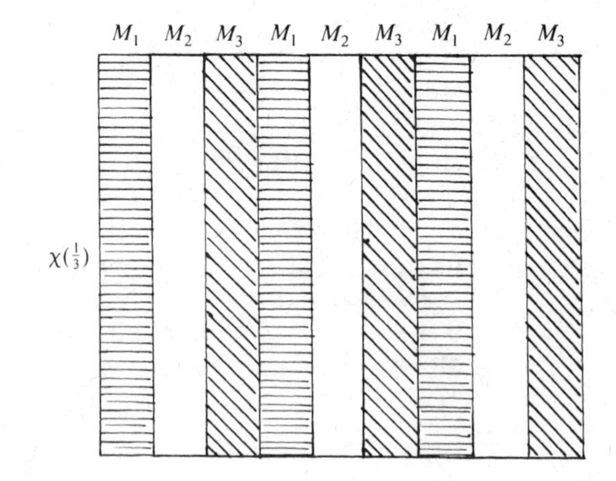

(b)

FIG. 18.3

all having the density $\delta = \frac{1}{3}$, and so placed that they cover the entire plane R^2 with their respective colors $\gamma_1, \gamma_2, \gamma_3$. In the sequel, sets of three monochromes covering the plane will be of particular interest.

The perfect tool for dealing with monochromes is the distance to the nearest integer $\{x\}$, which is defined by

$$\{x\} = \min_{m \in \mathbb{Z}} |x - m|. \tag{3.3}$$

Its graph is shown in Figure 18.4 and it is related to the function (2.4) of Chapter 15 by the identity $\{x\} = \langle 2x \rangle / 2$.

Let $\lambda^1 u_1 + \lambda^2 u_2 + a$ denote a linear function of u_1 and u_2, where

$$(\lambda^1)^2 + (\lambda^2)^2 > 0.$$

I claim:

The single inequality

$$\{\lambda^1 u_1 + \lambda^2 u_2 + a\} \leqslant \frac{\delta}{2} \qquad (0 < \delta < 1), \tag{3.4}$$

defines in the plane a set of points which is a monochrome $M(\delta)$ of density δ.

To see this we must realize that the inequality (3.4) is equivalent to the infinite system of inequalities

$$j - \frac{\delta}{2} \leqslant \lambda^1 u_1 + \lambda^2 u_2 + a \leqslant j + \frac{\delta}{2} \qquad (\text{with } j \in \mathbb{Z}), \tag{3.5}$$

because (3.4) says that the real number $\lambda^1 u_1 + \lambda^2 u_2 + a$ differs from some integer j by no more than $\delta/2$ in absolute value. However, the inequalities (3.5) define an MC $M(\delta)$ of density δ, with period and width given by

$$p = \frac{1}{\sqrt{(\lambda^1)^2 + (\lambda^2)^2}}, \qquad w = \frac{\delta}{\sqrt{(\lambda^1)^2 + (\lambda^2)^2}}. \tag{3.6}$$

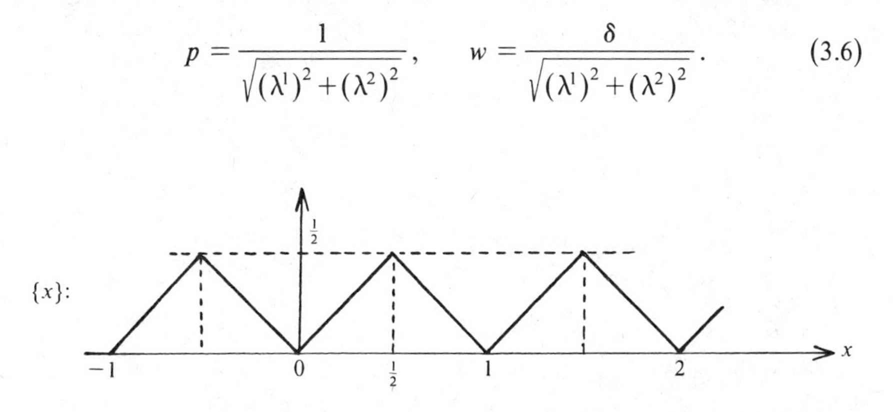

FIG. 18.4

Let us now consider two monochromes

$$\{\lambda_1^1 u_1 + \lambda_1^2 u_2 + a_1\} \leqslant \frac{\delta}{2},$$

$$\{\lambda_2^1 u_1 + \lambda_2^2 u_2 + a_2\} \leqslant \frac{\delta}{2}, \tag{3.7}$$

both having the same density δ, and such that

$$\begin{vmatrix} \lambda_1^1 & \lambda_1^2 \\ \lambda_2^1 & \lambda_2^2 \end{vmatrix} \neq 0. \tag{3.8}$$

We shall regard two such pairs of monochromes as equivalent, provided that one arises from the other by a nonsingular affine transformation of the variables. One such transformation is

$$u_1' = \lambda_1^1 u_1 + \lambda_1^2 u_1 + a_1,$$

$$u_2' = \lambda_2^1 u_1 + \lambda_2^2 u_2 + a_2, \tag{3.9}$$

and it transforms the monochromes (3.7) into the monochromes

$$M_1(\delta):\ \{u_1\} \leqslant \frac{\delta}{2},$$

$$M_2(\delta):\ \{u_2\} \leqslant \frac{\delta}{2}, \tag{3.10}$$

where we have omitted the primes on the variables. We see that for a given density δ, with $0<\delta<1$, there is essentially only *one* such pair of monochromes, as given by the inequalities (3.10). It should be equally clear that the pair of monochromes (3.10) do not cover the entire plane R^2 with their respective colors γ_1 and γ_2. Rather a lattice of infinitely many squares of dimensions $(1-\delta)\times(1-\delta)$ are left uncovered (please draw a diagram!).

For this reason we consider *three monochromes*

$$M_1(\delta):\ \{\lambda_1^1 u_1 + \lambda_1^2 u_2 + a_1\} \leqslant \frac{\delta}{2}, \tag{3.11}$$

$$M_2(\delta):\ \{\lambda_2^1 u_1 + \lambda_2^2 u_2 + a_2\} \leqslant \frac{\delta}{2}, \tag{3.12}$$

$$M_3(\delta):\ \{\lambda_3^1 u_1 + \lambda_3^2 u_2 + a_3\} \leqslant \frac{\delta}{2}, \tag{3.13}$$

and introduce the following definitions.

DEFINITION 2. *We say that the three monochromes* (3.11),(3.12),(3.13) *form a 3-chromo*

$$\chi(\delta) = (M_1(\delta), M_2(\delta), M_3(\delta)), \tag{3.14}$$

provided that they cover the entire plane, hence

$$M_1(\delta) \cup M_2(\delta) \cup M_3(\delta) = R^2. \tag{3.15}$$

DEFINITION 3. *The* 3-*chromo* (3.14) *is said to be in general position* (shortened to G.P.), *provided that no two among the monochromes $M_\nu(\delta)$ are parallel.*

Clearly, the conditions that $\chi(\delta)$ should be in general position are the inequalities

$$\begin{vmatrix} \lambda_1^1 & \lambda_1^2 \\ \lambda_2^1 & \lambda_2^2 \end{vmatrix} \neq 0, \quad \begin{vmatrix} \lambda_1^1 & \lambda_1^2 \\ \lambda_3^1 & \lambda_3^2 \end{vmatrix} \neq 0, \quad \begin{vmatrix} \lambda_2^1 & \lambda_2^2 \\ \lambda_3^1 & \lambda_3^2 \end{vmatrix} \neq 0. \tag{3.16}$$

A. *Examples.* We have already exhibited a 3-chromo $\chi(\frac{1}{3})$ in Figure 18.3(b). Indeed, the three monochromes (3.2) evidently cover the plane (Definition 2). However, notice that $\chi(\frac{1}{3})$ is not in G.P., because any two among the (3.2) are parallel.

Our second example is fundamental in the sequel. Its monochromes are all of density $\delta = 2/3$, and are defined by the inequalities

$$\begin{aligned} M_1(\tfrac{2}{3})&: \ \{u_1\} \leqslant \tfrac{1}{3}, \\ M_2(\tfrac{2}{3})&: \ \{u_2\} \leqslant \tfrac{1}{3}, \\ M_3(\tfrac{2}{3})&: \ \{u_1 + u_2\} \leqslant \tfrac{1}{3}. \end{aligned} \tag{3.17}$$

The 3-chromo

$$\chi^*(\tfrac{2}{3}) = (M_1(\tfrac{2}{3}), M_2(\tfrac{2}{3}), M_3(\tfrac{2}{3})) \tag{3.18}$$

is shown in Figure 18.5. The vertical and horizontal strips of M_1 and M_2 are seen to cover R^2 with the exception of a lattice of squares

$$s_{ij} \text{ with center } (\tfrac{1}{2} + i, \tfrac{1}{2} + j) \qquad (i, j \in \mathbb{Z}) \quad \text{and sides} = \tfrac{1}{3}.$$

All s_{ij} are neatly covered by the third MC, $M_3(\frac{2}{3})$, for the following reason: The two boundary lines of its strip

$$-\tfrac{1}{3} \leqslant u_1 + u_2 \leqslant \tfrac{1}{3} \tag{3.19}$$

are seen to pass through the vertices $(\frac{1}{3}, -\frac{2}{3})$, $(\frac{2}{3}, -\frac{1}{3})$ of the square $S_{0,-1}$. Since the slanting MC of Figure 18.5 has visibly the density $\frac{2}{3}$, while (3.19) is one of its strips, we conclude that $M_3(\frac{2}{3})$ is correctly shown in Figure 18.5.

B. *An extremum property of the* 3*-chromo* $\chi^*(\frac{2}{3})$. In §4 we establish

THEOREM 1′. *Among all* 3*-chromos* $\chi(\delta)$ *which are in general position, the* 3*-chromo* $\chi^*(\frac{2}{3})$, *of* Figure 18.5, *has the least density*

$$\delta^* = \tfrac{2}{3}. \tag{3.20}$$

We may express Theorem 1′ equivalently as follows:
If we define

$$\delta^* = \inf\{\delta;\ \text{for all densities } \delta \text{ of 3-chromos in G.P.}\}, \tag{3.21}$$

then (3.20) *holds, and* $\delta^* = \frac{2}{3}$ *is the density of the* 3*-chromo* (3.18).

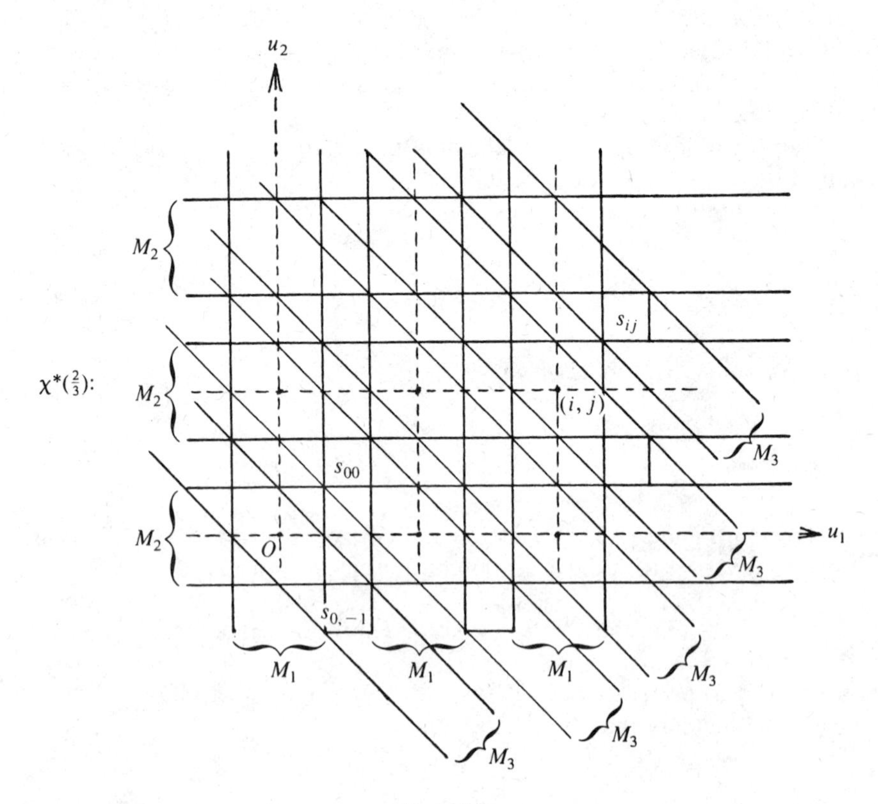

FIG. 18.5

We will also show *that* (3.18), *or an affine image of it, is the only* 3-*chromo in general position having the least density* δ^*.

In §5 we will show that Theorem 1′ is equivalent to Theorem 1 of our Introduction.

REMARKS. 1. We have shown in Figure 18.3(b) that the monochromes (3.2) form a 3-chromo $\chi(\frac{1}{3})$ of density $\delta = \frac{1}{3}$, hence $\delta = \frac{1}{3} < \delta^* = \frac{2}{3}$. *Does this contradict our Theorem* 1′? Not at all, because $\chi(\frac{1}{3})$ is not in G.P. as already mentioned in our first Example.

2. The definition (3.21) of δ^* and the 3-chromo $\chi^*(\frac{2}{3})$ demonstrate the inequality

$$\delta^* \leqslant \tfrac{2}{3}. \tag{3.22}$$

4. A Proof of Theorem 1′. We will actually prove Theorem 1′ in the following sharper form.

THEOREM 1″. *Let*

$$\chi(\delta) = (M_1(\delta), M_2(\delta), M_3(\delta)) \tag{4.1}$$

be a 3-*chromo in general position. The assumption that*

$$\delta \leqslant \tfrac{2}{3} \tag{4.2}$$

implies that

$$\delta = \tfrac{2}{3} \tag{4.3}$$

and that $\chi(\delta)$ *is equivalent with the special* 3-*chromo* $\chi^*(\frac{2}{3})$ *of Figure* 18.5.

We already know that we lose no generality by choosing the first two monochromes of (4.1) as shown in Figure 18.6, since this can be achieved by an appropriate nonsingular affine transformation. Writing

$$\sigma = 1 - \delta, \tag{4.4}$$

we notice that the union $M_1(\delta) \cup M_2(\delta)$ already covers the plane with the exception of the lattice of open squares

$$s(m_1, m_2) = \{m_1 < u_1 < m_1 + \sigma, m_2 < u_2 < m_2 + \sigma; (m_1, m_2) \in \mathbb{Z}^2\}. \tag{4.5}$$

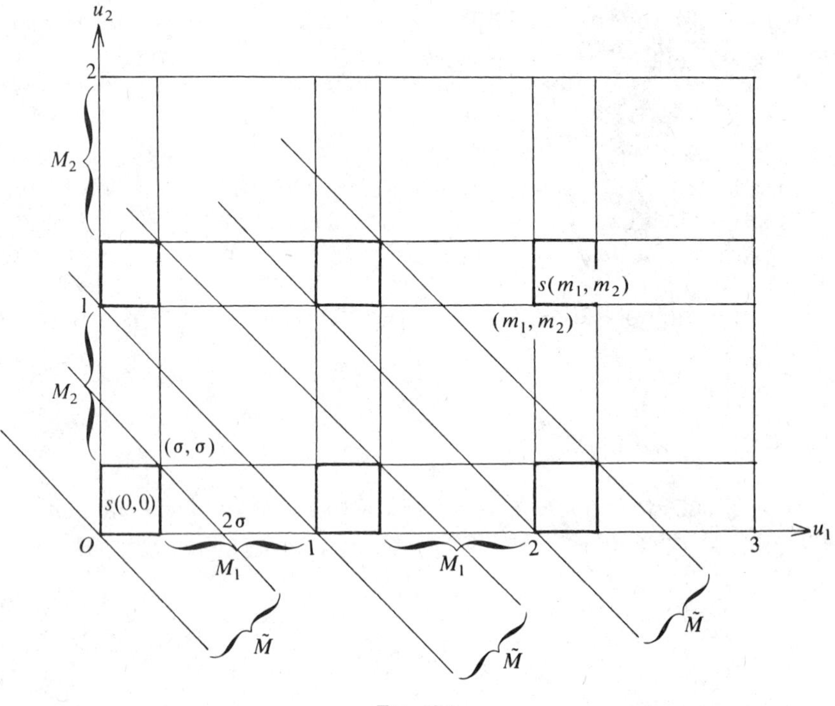

FIG. 18.6

We form the union of all these squares

$$\Sigma = \bigcup_{(m_1, m_2)} s(m_1, m_2). \tag{4.6}$$

Because (4.1) is a 3-chromo (Definition 2 of §3), the third $M_3(\delta)$ is to cover all these squares, so that we must have

$$M_3(\delta) \supset \Sigma. \tag{4.7}$$

We do not have much control over the location of $M_3(\delta)$, but we can certainly assume that $M_3(\delta)$ should have its strips parallel to a line

$$u_1 + \gamma u_2 = 0 \tag{4.8}$$

of *negative* slope. For, if not, then a reflection of the figure in the vertical line $u_1 = \sigma/2$ would achieve this purpose, while (4.7) would still hold. We may therefore assume that in (4.8) we have

$$\gamma > 0. \tag{4.9}$$

We shall now state several propositions, call them Lemmas if you like.

I. *The assumption* (4.7) *implies that*

$$\gamma = 1. \tag{4.10}$$

Proof: We assume that (4.10) does not hold, hence that

$$\gamma \neq 1, \tag{4.11}$$

and reach a contradiction. For this purpose we project all the squares (4.5) onto the u_1-axis in a slanting direction parallel to the line (4.8). The equations of the lines parallel to (4.8) and passing through its corners (m_1, m_2) and $(m_1+\sigma, m_2+\sigma)$, are

$$u_1 - m_1 + \gamma(u_2 - m_2) = 0 \quad \text{and} \quad u_1 - m_1 - \sigma + \gamma(u_2 - m_2 - \sigma) = 0,$$

respectively. Their intersections with the line $u_2 = 0$ are the points

$$u_1 = m_1 + \gamma m_2 \quad \text{and} \quad u_1 = m_1 + \sigma + \gamma(m_2 + \sigma).$$

It follows that the open square $s(m_1, m_2)$ is projected, parallel to (4.8), onto the open interval

$$I(m_1, m_2) = (m_1 + \gamma m_2, m_1 + \gamma m_2 + \sigma(1+\gamma)) \tag{4.12}$$

of the u_1-axes $\mathbb{R}^1$. But then, the projection of the entire set (4.6) is the (open) set

$$\Omega = \cup I(m_1, m_2). \tag{4.13}$$

We distinguish two cases.

1. γ *is irrational.* By the one-dimensional Kronecker Theorem of Chapter 16, §2, we conclude that the points $m_1 + \gamma m_2$, which are the left endpoints of the intervals (4.12), are dense on $\mathbb{R}^1$ (see also [**1**, §23.1]). It follows that the intervals (4.12), being of fixed length $\sigma(1+\gamma)$, cover the entire axis, hence

$$\Omega = \mathbb{R}^1. \tag{4.14}$$

However, our assumption (4.7) surely implies that $M_3(\delta) \supset \Omega$, and now (4.14) shows that

$$M_3(\delta) \supset \mathbb{R}^1. \tag{4.15}$$

This conclusion is clearly impossible, because the intersection $M_3(\delta)\cap\mathbb{R}^1$ is composed of a sequence of equidistant closed and disjoint intervals of equal lengths. *Reason*: $M_3(\delta)$ is a monochrome of density $\delta\leqslant\frac{2}{3}$.

2. γ *is rational.* Let

$$\gamma=\frac{a}{b},\quad\text{where}\quad(a,b)=1. \tag{4.16}$$

By our assumption (4.11) we must have $a+b\geqslant 3$, and therefore

$$\sigma(1+\gamma)=\sigma\frac{a+b}{b}\geqslant\frac{3\sigma}{b}. \tag{4.17}$$

Moreover

$$m_1+\gamma m_2=m_1+\frac{a}{b}m_2=\frac{bm_1+am_2}{b}=\frac{j}{b}, \tag{4.18}$$

where the numerator j may assume any integer value. This means: *The left endpoints of the intervals* (4.12) *form an infinite arithmetic progression of step* $1/b$.

II. *Under our assumption* (4.2) *it is impossible that*

$$\delta<\tfrac{2}{3}. \tag{4.19}$$

Proof: Let us assume that (4.19) holds and reaches a contradiction. From (4.19) we conclude that $\sigma=1-\delta>1-\frac{2}{3}=\frac{1}{3}$, and therefore

$$3\sigma>1. \tag{4.20}$$

Now (4.17) and (4.20) show that

$$\sigma(1+\gamma)>\frac{1}{b}. \tag{4.21}$$

From (4.18) and (4.21) we conclude that our *open intervals* (4.12) *actually overlap and must therefore cover the entire axis.* So again we must have (4.14), leading to the impossible conclusion (4.15). We have just shown that (4.10) holds.

We have also just shown that the assumption (4.2) of Theorem 1″ necessarily implies its conclusion (4.3). From $\delta=\frac{2}{3}$ we obtain that $\sigma=1-\delta=\frac{1}{3}$ and our Figures 18.5 and 18.6 become identical if we shift the origin of the latter to the point $(\frac{1}{3},\frac{1}{3})$. It follows that the 3-chromos $\chi(\delta)$ and $\chi^*(\frac{2}{3})$ are identical.

5. The Connection Between K-S Polyhedra and 3-Chromos. In §2 we considered a K-S polyhedron

$$\Pi\colon x_\nu = \langle \lambda_\nu^1 u_1 + \lambda_\nu^2 u_2 + a_\nu \rangle \qquad (\nu = 1,2,3), \tag{5.1}$$

within the unit cube γ_3, and said that it is ρ-admissible, provided that Π does not penetrate into the open cube

$$C_\rho\colon \ \|x - c\|_\infty < \rho. \tag{5.2}$$

In §3 we studied a 3-chromo of density δ

$$\chi(\delta) = (M_1(\delta), M_2(\delta), M_3(\delta)) \tag{5.3}$$

composed of the three monochromes

$$M_\nu(\delta)\colon \ \{\lambda_\nu^1 u_1 + \lambda_\nu^2 u_2 + a_\nu\} \leqslant \frac{\delta}{2} \qquad (\nu = 1,2,3). \tag{5.4}$$

Notice that the same terms λ_ν^1, λ_ν^2, a_ν are used in (5.1) and (5.4). Moreover, we assume that the "radius" ρ and the density δ satisfy the equation

$$\delta + 2\rho = 1 \qquad (0 < \delta < 1, 0 < \rho < \tfrac{1}{2}). \tag{5.5}$$

Our aim is to establish the following *equivalence theorem.*

THEOREM 2. *If* (5.5) *holds, then*

$$\textit{the K-S polyhedron } \Pi \textit{ is } \rho\textit{-admissible} \tag{5.6}$$

if and only if

$$\textit{the 3-chromo } \chi(\delta) \textit{ has the density } \delta. \tag{5.7}$$

Proof: Let

$$B_\rho = \gamma_3 \backslash C_\rho \qquad \text{(Figure 18.7(b))}, \tag{5.8}$$

denote the closed cubical shell that we obtain by removing the open cube C_ρ from inside γ_3. Then (5.6), i.e., the ρ-admissibility of Π, means that

$$\Pi \subset B_\rho. \tag{5.9}$$

We now use properties of the Euler spline $\langle x \rangle$ having the graph of Figure

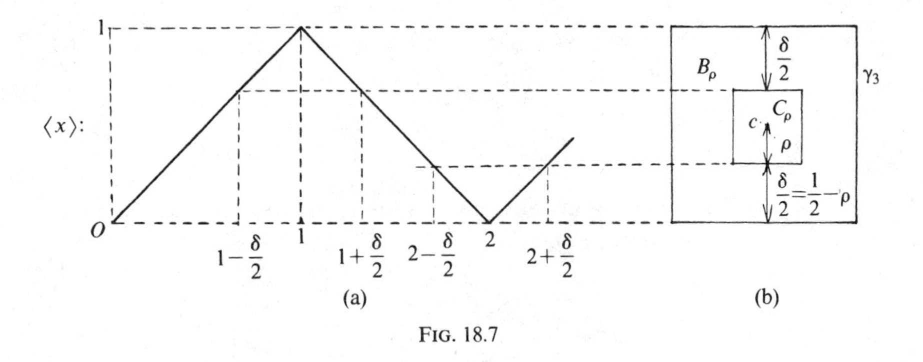

FIG. 18.7

18.7(a). An equivalent way of expressing (5.9) is as follows:

For any (u_1, u_2) *some one of the three numbers* $\lambda^1_\nu u_1 + \lambda^2_\nu u_2 + a_\nu$ $(\nu = 1,2,3)$ *differs from some integer by no more than* $\frac{1}{2} - \rho$ *in absolute value.* (5.10)

However, by (5.5) we have that $\frac{1}{2} - \rho = \delta/2$ (Figure 18.7(b)). Therefore we may restate (5.10), equivalently, as follows:

For any (u_1, u_1) *some one of the three numbers* $\lambda^1_\nu u_1 + \lambda^2_\nu u_2 + a_\nu$ $(\nu = 1,2,3)$ *differs from some integer by no more than* $\dfrac{\delta}{2}$ *in absolute value.* (5.11)

Using now the definition (3.3) of the function $\{x\}$, we may reformulate (5.11), equivalently, as follows:

For any (u_1, u_2) *some one of the three numbers* $\{\lambda^1_\nu u_1 + \lambda^2_\nu u_2 + a_\nu\}$ $(\nu = 1,2,3)$ *does not exceed* $\delta/2$. (5.12)

If we recall the monochromes (5.4), we may reformulate (5.12), equivalently, by the set inclusion

$$M_1(\delta) \cup M_2(\delta) \cup M_3(\delta) \supset R^2. \tag{5.13}$$

However, this last condition (5.13) states that (5.3) is a 3-chromo.

We have just proved the "only if" part of Theorem 2, hence that (5.6) $\Rightarrow$ (5.7). However, the steps can be reversed to obtain:

$$(5.7) \Rightarrow (5.13) \Rightarrow (5.12) \Rightarrow (5.11) \Rightarrow (5.10) \Rightarrow (5.9) \Rightarrow (5.6),$$

or (5.7) $\Rightarrow$ (5.6), and so (5.6) $\Leftrightarrow$ (5.7), proving Theorem 2.

In Definition 1 of §2 we have defined what we mean that (5.1) should be in general position. Likewise, in Definition 3 of §3 we have defined when the 3-chromo (5.3) is in general position. However, from (2.2) and (3.16) we see that both definitions amount to the same conditions on the 3×2 matrix $\|\lambda^i_\nu\|$: that its three second order minors be $\neq 0$. We state this as

COROLLARY 1. *The* K-S *polyhedron* (5.1) *is in general position if and only if the* 3-*chromo* (5.3) *is in general position.*

6. A Proof of Theorem 1. The above correspondence between ρ-admissible K-S polyhedra Π in general position, and 3-chromos $\chi(\delta)$ of density δ in general position, set up by the equation

$$\delta + 2\rho = 1, \tag{6.1}$$

allows us to derive immediately a proof of Theorem 1. It is clear that if we wish to *maximize* ρ, and this is what Theorem 1 is about, we must *minimize* δ. It follows from (6.1) that the extreme values ρ^* and δ^*, defined by (2.10) and (3.21), are connected by the equation

$$\delta^* + 2\rho^* = 1. \tag{6.2}$$

However, we have shown in §4 that

$$\delta^* = \tfrac{2}{3}, \tag{6.3}$$

and that this least value is reached essentially only by the 3-chromo $\chi^*(\frac{2}{3})$ defined by the equations (3.17) and shown in Figure 18.5. It follows from (6.2) and (6.3) that

$$\rho^* = \tfrac{1}{2} - \tfrac{1}{2}\delta^* = \tfrac{1}{2} - \tfrac{1}{3} = \tfrac{1}{6}, \tag{6.4}$$

and that this largest value is reached essentially only by the K-S polyhedron

$$\Pi^*: \begin{array}{l} x_1 = \langle u_1 \rangle, \\ x_2 = \langle u_2 \rangle, \\ x_3 = \langle u_1 + u_2 \rangle. \end{array} \tag{6.5}$$

This is identical with Kepler's tetrahedron $T' = A'B'C'D'$ of (2.7), and shown in Figure 18.1. The first tetrahedron $T = ABCD$ of (2.6), leads to the same $\chi^*(\frac{2}{3})$ expressed slightly differently. This establishes Theorem 1.

7. A Brief Report on Generalizations of the Problem. In [4] the extremum problem just solved is generalized as follows. A k-dimensional flat

$$L^k_n\colon\ x_\nu = \sum_{i=1}^{k} \lambda^i_\nu u_i + a_\nu \qquad (\nu = 1, \ldots, n) \qquad (1 \leqslant k < n),$$

is reflected by the $2n$ facets of the cube γ_n of the space R^n, giving rise to a K-S polytope

$$\Pi_n^k: \; x_\nu = \left\langle \sum_{i=1}^{k} \lambda_\nu^i u_i + a_\nu \right\rangle \qquad (\nu=1,\ldots,n). \tag{7.1}$$

This polytope is said to be ρ-*admissible* $(0<\rho<\frac{1}{2})$, provided that Π_n^k does not penetrate into the open cube

$$C_\rho: \; \|x-c\| \leqslant \rho, \quad \text{where } x=(x_1,\ldots,x_n), \quad c=(\tfrac{1}{2},\ldots,\tfrac{1}{2}). \tag{7.2}$$

Π_n^k is said to be *in general position*, provided that the $n\times k$ matrix $\|\lambda^i\|$ has no vanishing minor of order k. For all polytopes Π_n^k satisfying these conditions, we define the quantity

$$\rho_{k,n}^* = \sup \rho. \tag{7.3}$$

The problem is to determine, or estimate, this quantity, and to determine the Π_n^k *which are* $\rho_{k,n}^*$*-admissible.*

Some of the results so far obtained are the following:

(i) In [4] it is shown that

$$\rho_{k,n}^* \geqslant \frac{1}{2} - \frac{k}{2n} \qquad (1 \leqslant k \leqslant n-1).$$

(ii) Also in [4] I conjecture that we have here the equality sign

$$\rho_{k,n}^* = \frac{1}{2} - \frac{k}{2n} \qquad (1 \leqslant k \leqslant n-1).$$

(iii) This conjecture is established in [4] for the two extreme values of k: $k=1$ and $k=n-1$.

(iv) The paper [5] studies the structure of Π_n^{n-1}, which is $\rho_{n-1,n}^*$-admissible, where $\rho_{n-1,n}^* = 1/2n$. This polytope generalizes the Stella Octangula in R^n.

(v) Figure 15.3(a) and (b), which we reproduce in Figure 18.8 below, concerns the case when $k=1$ and $n=3$, when

$$\rho_{1,3}^* = \frac{1}{2} - \frac{1}{2\cdot 3} = \frac{1}{2} - \frac{1}{6} = \frac{1}{3}.$$

It shows the two extremizing K-S polygons, i.e., billiard ball motions. The first is a *hexagon*, the second an *octagon*, and both are seen to wind themselves around the maximal possible cube

$$C_{1/3}\colon \; \|x-c\|_\infty < \tfrac{1}{3},$$

in the sense that every one of their sides touches an edge of the cube. Figure 18.8 is borrowed from [**3**, page 278].

Problems

1. We deal with a single special case of the general problem of §7: the case when $k=1, n=3$. The problem is to show that

$$\rho^*_{1,3} = \tfrac{1}{3}.$$

Hint: The analogue of the equivalence Theorem 2 holds with unchanged proof: *We assume that* $\delta + 2\rho = 1$. *The* K-S *polygon, or billiard ball motion*

$$\Pi^1_3\colon \; x_1 = \langle u \rangle, \qquad x_2 = \langle \lambda_2 u + a_2 \rangle, \qquad x_3 = \langle \lambda_3 u + a_3 \rangle,$$

is ρ-*admissible iff the* (*one-dimensional*!) *monochromes*

$$M_1(\delta)\colon \; \{u\} \leqslant \frac{\delta}{2}, \quad M_2(\delta)\colon \; \{\lambda_2 u + a_3\} \leqslant \frac{\delta}{2}, \quad M_3(\delta)\colon \; \{\lambda_3 u + a_3\} \leqslant \frac{\delta}{2},$$

form a 3-*chromo, which means that*

$$M_1(\delta) \cup M_2(\delta) \cup M_3(\delta) = \mathbb{R}^1.$$

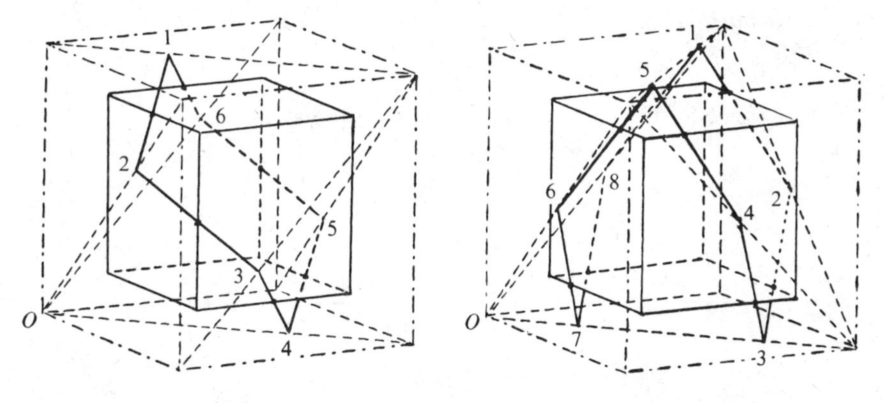

FIG. 18.8

It is clear that for this to hold, we must have that $\delta \geqslant \frac{1}{3}$, or else there would not be enough paint to cover the entire axis! A 3-chromo $\chi(\frac{1}{3})$ of density $\frac{1}{3}$ is shown in Figure 18.3(b). Therefore $\delta_{1,3}^{*} = \frac{1}{3}$.

Show that its monochromes admit the explicit representations

$$\chi(\tfrac{1}{3})\colon\ \{u\} \leqslant \tfrac{1}{6}, \qquad \{u-\tfrac{1}{3}\} \leqslant \tfrac{1}{6}, \qquad \{u-\tfrac{2}{3}\} \leqslant \tfrac{1}{6}.$$

From the equivalence theorem we conclude that

$$\Pi\colon\ x_1 = \langle u \rangle, \qquad x_2 = \langle u-\tfrac{1}{3}\rangle, \qquad x_3 = \langle u-\tfrac{2}{3}\rangle.$$

Since $\rho_{1,3}^{*} = \frac{1}{2} - \frac{1}{2}\delta_{1,3}^{*} = \frac{1}{2} - \frac{1}{6} = \frac{1}{3}$, we conclude that Π is $\frac{1}{3}$-admissible.

This, however, is not the only one: *In Figure* 18.9(b) *we show a second 3-chromo* $\tilde{\chi}(\frac{1}{3})$ *of density* $\frac{1}{3}$. Its intervals, i.e., its one-dimensional "strips", are alternatingly of lengths 1 and 2. The MC $M_1(\frac{1}{3})$ of Figure 18.9(b) is visibly of density $\frac{1}{3}$. Observe that the intervals of $M_2(\frac{1}{3})$ and $M_3(\frac{1}{3})$ occupy alternatingly *every other one* of the intervals of length 2. This gives the monochromes $M_2(\frac{1}{3})$ and $M_3(\frac{1}{3})$ the density $\delta = \frac{1}{3}$ automatically. $\chi(\frac{1}{3})$ and $\tilde{\chi}(\frac{1}{3})$ are the only one-dimensional 3-chromos of density $\delta = \frac{1}{3}$.

Check that its monochromes admit the representations

$$\tilde{\chi}(\tfrac{1}{3})\colon\ \{u\} \leqslant \tfrac{1}{6}, \qquad \{\tfrac{1}{2}u-\tfrac{1}{4}\} \leqslant \tfrac{1}{6}, \qquad \{\tfrac{1}{2}u-\tfrac{3}{4}\} \leqslant \tfrac{1}{6}.$$

Again using the equivalence theorem, we conclude that

$$\tilde{\Pi}\colon\ x_1 = \langle u \rangle, \qquad x_2 = \langle \tfrac{1}{2}u-\tfrac{1}{4}\rangle, \qquad x_3 = \langle \tfrac{1}{2}u-\tfrac{3}{4}\rangle$$

is a K-S polygon, which is also $\frac{1}{3}$-admissible. *Also check that* Π *and* $\tilde{\Pi}$ *are the hexagon and octagon of Figure* 18.8, *respectively, or equivalent to them by symmetries of* γ_3.

2. Show that for the case when $k=1$, $n=2$, we have

$$\rho_{1,2}^{*} = \tfrac{1}{4},$$

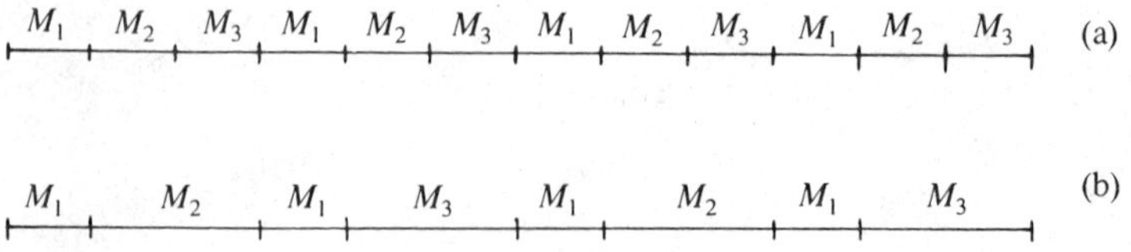

M_1 M_2 M_3 M_1 M_2 M_3 M_1 M_2 M_3 M_1 M_2 M_3 (a)

M_1 M_2 M_1 M_3 M_1 M_2 M_1 M_3 (b)

FIG. 18.9

and that the only billiard ball path which is $\frac{1}{4}$-admissible is the square of Figure 15.2(c) which has its vertices in the midpoints of the sides of γ_2.

References

[1] G. H. Hardy and E. M. Wright, *An Introduction to the Theory of Numbers*, 4th ed., Oxford, 1960.

[2] D. König and A. Szücs, *Mouvement d'un point abandonné à l'intérieur d'un cube*, Rend. Circ. Mat. Palermo, 36 (1913) 79–90.

[3] I. J. Schoenberg, *Extremum problems for the motions of a billiard ball* II. *The* L_∞ *norm*, Indag. Math., 38 (1976) 263–279.

[4] ——, *Extremum problems for the motions of a billiard ball* III. *The multidimensional case of König and Szücs*, Studia Sci. Math. Hungarica, 13 (1978) 53–78.

[5] ——, *Extremum problems for the motions of a billiard ball* IV. *A higher-dimensional analogue of Kepler's Stella Octangula*, Studia Sci. Math. Hungarica, 14 (1979) 273–292.

INDEX